Construction on Contaminated Sites

Construction on Contaminated Sites

J. Scott Lowe, P.E.
Theodore J. Trauner, Jr., P.E., P.P.

McGraw-Hill
New York San Francisco Washington, D.C. Auckland Bogotá
Caracas Lisbon London Madrid Mexico City Milan
Montreal New Delhi San Juan Singapore
Sydney Tokyo Toronto

Library of Congress Cataloging-in-Publication Data

Lowe, J. Scott.
    Construction on contaminated sites / J. Scott Lowe, Theodore J. Trauner, Jr.
      p.   cm.
    Includes index.
    ISBN 0-07-038878-4
    1. Construction on contaminated sites. 2. Building sites—Risk assessment. 3. Building—Superintendence. I. Trauner, Theodore J. II. Title.
TD1095.5.L69   1996
690'.22—dc20   96-33169
                                                              CIP

McGraw-Hill
A Division of The McGraw·Hill Companies

Copyright © 1997 by The McGraw-Hill Companies, Inc. All rights reserved. Printed in the United States of America. Except as permitted under the United States Copyright Act of 1976, no part of this publication may be reproduced or distributed in any form or by any means, or stored in a data base or retrieval system, without the prior written permission of the publisher.

1 2 3 4 5 6 7 8 9 0   DOC/DOC   9 0 1 0 9 8 7 6

ISBN 0-07-038878-4

The sponsoring editor for this book was Larry S. Hager, the editing supervisor was Stephen M. Smith, and the production supervisor was Donald F. Schmidt. It was set in Century Schoolbook by Ron Painter of McGraw-Hill's Professional Book Group composition unit.

Printed and bound by R. R. Donnelley & Sons Company.

McGraw-Hill books are available at special quantity discounts to use as premiums and sales promotions, or for use in corporate training programs. For more information, please write to the Director of Special Sales, McGraw-Hill, 11 West 19th Street, New York, NY 10011. Or contact your local bookstore.

This book is printed on acid-free paper.

Information contained in this work has been obtained by The McGraw-Hill Companies, Inc. ("McGraw-Hill") from sources believed to be reliable. However, neither McGraw-Hill nor its authors guarantee the accuracy or completeness of any information published herein and neither McGraw-Hill nor its authors shall be responsible for any errors, omissions, or damages arising out of use of this information. This work is published with the understanding that McGraw-Hill and its authors are supplying information but are not attempting to render engineering or other professional services. If such services are required, the assistance of an appropriate professional should be sought.

*This book is dedicated to our wives, Ann and Dariel,
whose support and encouragement made it possible.*

Contents

Preface   ix
Introduction   xiii

Chapter 1. Defining Contamination   1

    Asbestos   2
    Lead   3
    Underground Storage Tanks   4
    PCBs   4

Chapter 2. Preventing the Unexpected Discovery of Contamination   7

    The Risks   7
    Research   9
    Inspection   11
    Testing   14
    Interpretation   14
    Consultant Selection   15
    Other Project Participants   16
    Summary   19

Chapter 3. Preemptive Actions before the Project Begins   21

    Planning   21
    Execution   39
    Summary   66

Chapter 4. The Contract as a Risk Management Tool   67

    The Changes Clause   68
    Variation of Quantities   79
    Scheduling   80
    Suspensions of Work   84
    Liquidated Damages and Time Extension Clauses   85

viii Contents

Termination	89
Dispute Prevention and Resolution	95

Chapter 5. Immediate Actions upon Discovery of Contamination 101

Discovery	101
Notification	103
Safety and Health Response	105
Containment	111
Evaluation	112
Documentation	113
Summary	119

Chapter 6. Measuring the Time Impact of the Discovery of Contamination 121

Analyzing Delays Using a CPM Schedule	122
Analyzing Delays Using a Bar Chart	127
The Impacted As-Planned Approach	129
Collapsed As-Built Analysis	131
Contemporaneous Analysis Approach	131
Acceleration	137

Chapter 7. The Impact of the Discovery of Contamination on Efficiency 141

The Measured Mile	145
When the Measured Mile Cannot Be Used	146
The Total Cost Approach	147
Acceleration	148

Chapter 8. Measuring the Costs of Discovering Contamination 151

Estimated versus Actual Costs	151
Labor Costs Effects	152
Equipment Cost Effects	157
Material Cost Effects	164
Overhead and Other Markups	165

Chapter 9. Dealing with Disputes Arising from the Discovery of Contamination 173

Nonbinding Dispute Resolution Techniques	173
Binding Arbitration, Courts, and Contract Appeals Boards	181

Appendix A Requirements for Owners and Operators of Underground Storage Tanks 183
Appendix B Special Contractor Prequalification Questionnaire 195
Appendix C OSHA Regulations: Hazard Communications Standard 205
Appendix D Action Items Control Log 227
Appendix E Sample Written Hazard Communication Program 231
Appendix F Excerpt from *Modification Impact Evaluation Guide* 237
Index 251

Preface

Construction on a contaminated site is a possibility that faces many in the construction industry. There are a number of reasons for this. First, many of the most desirable sites in America's urban areas have already been developed. Those that remain undeveloped often possess some defect that makes them less desirable. One common defect is the presence of contamination. As more developers are forced to choose these less desirable sites, the possibility increases that the construction that follows will be on a contaminated site. Also, the remodeling and renovation of existing structures is a popular alternative to new construction. Existing structures, however, were often constructed using materials that are no longer considered safe. Asbestos is probably the most common example.

A third source of contaminated projects may come from the efforts of those trying to revive economically depressed areas and regions. In some cases, these areas possess extensive industrial and commercial resources. These resources, however, sit abandoned or underutilized. At least one reason for this idleness appears to be that these sites are or may be contaminated. The risks and costs associated with this contamination scare away potential investors. If those interested in encouraging development of these areas, sometimes known collectively as *brownfields*, are successful in reducing investors' fears, then the number of projects involving these potentially contaminated sites should increase.

Those that are or may become involved with construction on contaminated sites have three primary objectives. Their first objective is to do everything possible to avoid encountering contamination in the first place. The second objective is to plan and execute the project work in such a way as to minimize the impact of the presence or discovery of contamination. The third objective is to be able to mitigate and measure the impacts of the discovery of contamination on the construction site. This book has been designed to address each of these three primary objectives. It has also been designed to help construction project participants reduce, manage, or, where possible, eliminate the risk associated with contamination on construction sites.

Chapter 1 presents a simple definition of contamination. There is also a discussion of typical contaminants and where they are likely to be found. This chapter also contains a brief discussion of the regulatory framework within which construction on contaminated sites is bound.

Chapter 2 discusses the ways that owners and developers—and the architects, engineers, contractors, and others that assist them—can avoid becoming involved in a project where the discovery of contamination is a possibility. The methods that are discussed to accomplish this objective include approaches to researching, inspecting, testing, and evaluating potential projects and project sites. Keep in mind that while these investigative efforts may make the avoidance of contaminated sites possible, this is not a risk that can be completely removed.

Chapter 3 is a thorough discussion of the use of project management techniques to reduce the impacts of the discovery of contamination on the project. This includes a discussion of the planning steps to take. This chapter also addresses steps in the execution of the management plan that can reduce the impact of discovery. Discussions in Chap. 3 range from the successful use of partnering to proper scheduling techniques.

Chapter 4 identifies, describes, and provides examples of contract clauses that are essential to addressing the risks associated with the discovery of contamination. This includes a discussion of changes clauses, the differing site condition clause, and the variation in quantity clause. Because the potential for disputes is greater on contaminated sites, particularly on sites where contamination is discovered unexpectedly, this chapter also discusses the contract language available to help prevent, reduce, and resolve disputes on construction projects, including a discussion of bid document escrow and dispute review boards. In conjunction with Chap. 3, Chap. 4 also explores some of the basic contract forms such as lump-sum, cost-plus, and unit-price contracting, and identifies when the use of these contract formats is appropriate.

Chapter 5 presents a discussion of the immediate actions that all parties should consider taking when contamination is discovered. These actions range from health and safety considerations to notification. Proper formats for documentation are presented and discussed as well. Applicable regulations are discussed in detail.

With the discovery of contamination on the site, attention will naturally shift to determining and mitigating the effects of this discovery on the project. Potential effects range from delays to added costs and reductions in the efficiency of performing the project work. Chapter 6 describes how to identify and measure the potential time effects of discovery. Chapter 7 addresses potential effects on the contractor's work efficiency. Chapter 8 addresses the cost considerations associated with the discovery of contamination.

Because the discovery of contamination can lead to disputes and litigation, Chap. 9 discusses the many avenues available for the resolution of disputes on the project. Approaches discussed include negotiation, mediation, arbitration, and litigation. Also discussed are the use of disputes review boards and other forms of alternative disputes resolution techniques.

The risk that contamination will be discovered on the construction project is a risk best avoided. If it cannot be avoided, it is a risk that can be reduced and managed. Risk reduction and management are the result of the application of good project management techniques and skills.

Acknowledgments

The authors would like to acknowledge the following individuals for their capable assistance: Frances Swiger, who spent innumerable hours typing, editing, indexing, and working on the myriad of little details required to assemble a book; Donna Lane, who developed the graphics; Stephanie Trauner, who reviewed this book and prepared the index; Bill Manginelli, whose dedicated efforts allowed the authors to focus on this book; and Rocco Vespe, Lee Marshall, Sheldon Weaver, David Starfield, and Timothy Sullivan, whose careful research and thoughtful comments were essential.

J. Scott Lowe, P.E.
Theodore J. Trauner, Jr., P.E., P.P.

Introduction

The decaying hulks of America's commercial and industrial past occupy thousands of acres of its urban and rural landscape. Though they sit on valuable land, in choice locations, and in communities that desperately need the jobs and income that come with new development, they sit idle, untouched, and unwanted. These areas are known as *brownfields*. They sit idle while *greenfields* absorb the bulk of new commercial, residential, and industrial development.

Rather than renovating and reusing existing structures and facilities, greenfield development requires the construction of new buildings, new houses, and new parking lots. Also, because this new development takes place in previously undeveloped areas, it requires huge investments in new infrastructure. These investments are essentially duplicitive, since the infrastructure already exists in the brownfield areas, sitting idle and underused like the properties it serves.

Though the brownfields are ignored for a number of reasons, a primary contributor to their idleness is the perception that they are contaminated with any number of dangerous substances and that cleaning up the contamination will require the expenditure of huge sums. Because the presence and extent of the contamination is not always known, the cost to clean up a brownfields property is often hard to estimate. Potential buyers shun such risky properties and, consequently, the brownfields properties sit idle.

Recognizing the potential of many brownfields properties, those interested in the development of these areas have begun to explore ways to manage and reduce both the risk and the costs associated with the purchase of brownfield properties. If these efforts are successful, it is probable that more in the construction industry will become involved with projects where contaminants are known to exist or where there is a high probability that they will be discovered.

The increased potential for brownfields development is not the only reason that those involved with construction are likely to see an increase in the number of projects where the existence or discovery of some kind of contamination is an issue. Every year, the pool of desirable, undeveloped building sites gets smaller. Consequently, the likelihood that a developer's, architect's, engineer's, or contractor's next project will involve the reuse or renovation of existing buildings, or new construction on a previously developed site, increases. Since the potential always exists that these sites may be contaminated, the potential that those involved with construction will see more contamination also increases. In summary, the potential for the discovery or existence of contamination on the project site is no longer a risk that can be easily avoided.

Most of the participants in a construction project do not actually have to deal with the contaminant itself. Cleanup, abatement, or remediation of site contamination is generally a task handled by specialists, who mobilize on the site to perform their work and then leave as soon as their work is completed. The fact that most people on a construction site do not actually have to deal with the contaminant itself does not mean they are unaffected by it. The opposite is true. The discovery of contamination or the presence of contamination in greater quantities than anticipated can wreak havoc on carefully prepared budgets and time schedules. Simultaneously with addressing these problems as they arise, project participants must also sort out responsibility for the added costs and lost time. In this kind of environment it is easy for working relationships to be strained and for the number and size of disputes to increase. The likelihood that a project might have to be abandoned or cut back significantly might also increase.

Given the potential for disaster, it is important for the participants in a project where the risk of discovering contamination exists to make every effort to reduce, manage, and, where possible, eliminate this risk. The primary objective of this book is to set out some of the ways in which risk on a potentially contaminated project can be reduced or managed. This objective is accomplished in a number of ways. The easiest way to avoid having to deal with contamination is to avoid project sites that are contaminated. If the potential for contamination cannot be avoided, then the next step is to plan the project so that the impact of the discovery of contamination is minimized. Typically this is accomplished through careful preparation, including the drafting of appropriate contract documents that address the potential risks.

The risk that contamination will be discovered cannot be eliminated. Thus, any discussion of contamination risk management must address the actions to be taken when it is discovered, both immediately and in the long term. It is also important that the parties to a project where contamination has been encountered know how to assess the effects. This includes both the monetary and time effects that might be associated with such a discovery. Participants must be able to calculate the added costs of the discovery. They also must be able to quantify and determine the responsibility for delays to the project that result from the discovery of contamination.

Often, the discovery of contamination will strain the project's working relationships—sometimes to the breaking point. Knowing this to be the case allows project participants to consider several dispute avoidance and resolution techniques. To take full advantage of these techniques, they must be addressed in the initial contracts among the parties, before the contamination is even discovered. Thus, minimizing the effects of the discovery of contamination requires careful consideration of the risks and action to assure that these risks are addressed in the contract.

Discovering dangerous substances on a construction site is a real risk. Through the careful application of some of the techniques discussed in this book, it is a risk that can be reduced and managed, though probably not eliminated.

Chapter 1

Defining Contamination

A *contaminant* is a toxic material found on the construction site in a place where it is not wanted. When a contaminant is found, it is a *hazardous material*. When it is disposed of, it is *hazardous waste*. There are hundreds of thousands of potential contaminants, from arsenic to zinc compounds. Some can cause illness or injury immediately; the effects of others may not be felt for decades. Some are quite common; others are extremely rare.

Contaminants can be found almost anywhere on a construction site. Asbestos, for example, was used as an insulating material, but asbestos fibers can also be found in flooring materials, exterior siding and roofing materials, and fireproofing. Pipe may be made of lead, but lead was a component of paints and solder as well. Moving underground, the list of potential contaminants is extensive. Underground storage tanks may have leaked their contents into the surrounding soils. Sites of former dry cleaners may be contaminated with dry-cleaning chemicals. PCBs, cadmium, and other less well-known chemicals have found their way to locations far separated from their original source or use. Hazardous materials have been placed in barrels and dumped, sometimes innocently, sometimes not so innocently. They have been poured into lagoons and have seeped slowly into the surrounding soils. Hazardous materials are now spread so widely in the environment that there is a potential that almost any construction project could be affected.

To protect the public from the threat to health and safety represented by hazardous materials, governments throughout the country have adopted strict laws, rules, and ordinances. In addition, they have spent billions of dollars on enforcement and litigation to make sure that these laws are obeyed. Many millions of dollars in fines have been assessed, and billions of dollars more have been spent on cleanup. At the federal level, a number of laws, agencies, and regulations govern the federal government's approach to the topic of hazardous materials and their handling and disposal. On the legislative side, the Comprehensive Environmental Response, Compensation, and Liability Act (CERCLA), the Resource Conservation and Recovery Act (RCRA), and a host of other lesser measures establish everything from the responsibility for the costs of cleanup to the disposal of hazardous wastes. Typically, these laws are administered by either the Occupational Safety and Health Administration (OSHA) or the Environmental Protection Agency (EPA). Most of the laws themselves do not address the details of enforcement with regard to each hazardous material. Instead, these details are left to the agencies to develop and issue as regulations. Title 40 of the *Code of Federal Regulations* includes most of these regulations. Over the years, Title 40 has expanded to more than 10,000 pages. Added to these federal requirements are a plethora of state and local requirements. The true definition of what constitutes contamination and what does not is defined within these many pages of regulations.

Following is a discussion of some of the contaminants that will most often concern those on construction projects. The list of contaminants discussed is not exhaustive, as the list of potentially hazardous materials would cover many pages. Rather, the objective is to touch on the most likely or probable contaminants that may be encountered.

Asbestos

Asbestos is perhaps the most common contaminant discovered on construction projects, particularly projects that involve the remodeling of, renovation of, or additions to existing buildings. Because it is resistant to heat, water, and microorganisms, is an effective insulator, resists wear, resists chemical attack, and can be added to make many products harder, smoother, or more opaque, asbestos was found in many building materials. Buildings constructed prior to the late 1970s often contain numerous asbestos-containing materials and products. It was used extensively as a thermal and acoustic insulation material. For these applications it was often sprayed or troweled on, or formed onto pipes, boilers, and heated tanks. Asbestos was also added to plaster for interior walls and ceilings, stucco, siding and roof shingles, floor and ceiling tiles, wallcoverings, and numerous other products.

Asbestos is a general term used to describe a group of inorganic hydrated silicates found in nature. Natural deposits or formations can be found throughout the United States and the world. The fact that there are some health risks associated with asbestos was probably recognized by the turn of the century. By the 1970s, its use had been reduced dramatically. Since that time there has been an active effort to remove it, particularly in its friable form. In this form, asbestos fibers can easily become airborne. It is these airborne fibers that represent the chief hazard of asbestos. They can become lodged in the lungs and lead to the development of asbestosis, cancer, and other breathing problems.

If it is discovered in an uncontained and friable form, asbestos usually must be removed. If it is contained or in a nonfriable form, it can often remain in place. Many qualified experts are capable of determining whether removal is necessary, and many contractors specialize in asbestos removal. Given the training, licensing, equipment, permitting, and disposal requirements, asbestos is a substance best handled by experts.

From a planning perspective, no work can take place in an area while asbestos removal is being accomplished. Generally, when removing friable materials, the removal area must be sealed off with the air pressure controlled to assure the capture of all escaped fibers. If a structure containing asbestos-bearing materials is to be demolished, even asbestos in a nonfriable state may very often have to be removed. Nonfriable asbestos is usually asbestos that cannot become easily airborne. Typically, it is asbestos that is contained within floor tiles and other products that hold the asbestos and prevent it from becoming airborne. During demolition, however, these materials can be broken and/or crushed, potentially releasing the asbestos fibers.

As with most contaminants, the best policy may be to avoid projects that will involve dealing with asbestos. When this is not possible, and it often is not, the best approach may be to remove the asbestos before construction begins. This approach is described in greater detail in Chap. 3.

If asbestos is encountered during construction or cannot be removed before construction begins, there are several approaches to its removal. A common approach is to phase the asbestos work in with other contract work. This approach is generally most appropriate in large renovation projects. In this scheme, asbestos removal is the first phase of work performed in an area after the area's occupants have been moved. Once the asbestos-removal phase is completed, the general construction work can begin.

Contractually, it is often best to provide for asbestos removal on a unit-price basis. Unit-price contracts and their application to projects where hazardous materials must be removed are discussed in Chaps. 3 and 4.

Lead

Lead may be a more prevalent contaminant than asbestos. It has been used in construction since ancient times. In structures constructed prior to 1978, it is possible that lead-based paints were used. Lead-based paints were used on interior walls, ceilings, window sills, window frames, doors, and door frames. Outside, lead-based paints were used on siding, trim, and fixtures.

Lead was also used to make pipe and was a common component of solder. In addition, because it was a common additive to gasoline, lead is also a widespread environmental contaminant, though unleaded gasolines have reduced this problem significantly.

The chief hazard represented by lead is through ingestion, typically as paint chips consumed by children, but also through drinking-water systems contaminated with lead. High concentrations of lead in the body can lead to mental retardation, blindness, and death.

Interestingly, until 1993 construction workers could be exposed to four times the amount of lead allowed by OSHA lead standards for other industries. The Housing and Community Development Act of 1992, however, reduced the exposure limits for construction workers by 75 percent, bringing them in line with other industries. This change probably has had its greatest effect on remodeling or renovation projects where paint removal is required.

Paint removal, particularly when abrasive techniques such as sandblasting and grinding are used, can cause lead to become airborne in dangerous concentrations. High airborne concentrations are significant because airborne lead is easily ingested. Failure to take adequate precautions when performing lead removal has resulted in the assessment of significant fines.

If lead-free project sites are not available, then project planners should try to remove the lead-containing paints before the start of construction. If this is also impossible, then it should be assumed that no other work activity can proceed in the area while lead-paint removal is taking place. Phasing the project to allow for lead-paint removal is a good approach to dealing with this problem.

Keep in mind that removal of lead-based paints is not always necessary. The need for removal is usually a function of the condition of the coating and the use of the facility.

Underground Storage Tanks

A significant environmental problem throughout the country is the existence of underground storage tanks (USTs). This problem is being addressed through a massive and expensive removal-and-replacement effort. It is doubtful, however, that every tank will be found, removed, and the contaminated soils properly disposed of. Consequently, the risk of discovering buried fuel tanks will continue.

A common reason for USTs is the storage of fuels. Fuels in the form of heating oil, gasoline, kerosene, and other similar liquids are good examples of contaminants that are contaminants only because they are in the wrong place. If they stay in their tanks and piping systems, they are not a contaminant. This is important to keep in mind because, unlike materials containing asbestos or lead-based paints, gasoline and other fuels will continue to be refined and used. They will also continue to be stored in buried tanks. Consequently, the possibility that buried tanks will be discovered on construction sites is a risk that will continue to exist. Admittedly, the impact of such a discovery should be reduced significantly by the latest generation of tanks, which are designed not to leak, but the risk of discovering buried tanks will continue.

Fuels also represent an area where the construction process itself can contaminate the site. Gasoline and diesel fuels are used extensively on construction sites. Spills and leaky storage containers can lead to contamination of a clean site.

As with most hazardous materials, if clean sites cannot be found, it is generally advisable to remove known buried tanks and contaminated soils before other construction work on the project begins.

PCBs

Polychlorinated biphenyls (PCBs) belong to a class of chemicals called chlorinated hydrocarbons. Their consistency ranges from oily liquids to waxy solids. PCBs have been put to a wide range of uses, but the majority of those produced have been used in the manufacture of electrical capacitors and transformers. A typical transformer might contain more than 200 gal of PCBs. They are useful in electrical components because of their insulating qualities. It is the destruction or improper disposal of electrical components that has caused the majority of PCB contamination.

PCBs present two major problems. First, they are toxic even at low concentrations and have been identified as potential carcinogens. They are also quite stable, remaining in the environment for years. Because they have been used in a wide range of products since the 1930s, and are so long lived, PCBs are widely distributed throughout the environment, with many potentially contaminated sites. Perhaps even more than the three contaminants discussed so far in this chapter, PCB cleanup is a task best left to professionals.

From a project planning perspective, avoiding project sites contaminated with PCBs is probably the best policy. If this is not possible, then fast-tracking or phased cleanup of PCBs is the preferred approach.

Asbestos, lead, USTs, and PCBs represent some of the more common contamination problems faced by those involved with construction on contaminated sites. There are, however, thousands of potential contaminants. Each has its own potential to disrupt a construction project. The next chapter discusses strategies for identifying these contaminants and their potential effects on project work.

Chapter

2

Preventing the Unexpected Discovery of Contamination

Given a choice, few if any construction owners would choose to become embroiled with contamination and its associated problems. True, some entities are in the business of cleanup, but for the majority of those who develop, build, expand, renovate, restore, design, or construct, contamination is best avoided. The reason for this is simple: Cleanup is costly. More than one entity has spent its entire construction budget cleaning up hazardous materials discovered on the project site. The objective of this chapter is to describe some techniques for controlling the risks associated with contamination being discovered on the project. Remember, however, that while it is possible to reduce and manage the risk of discovering contamination on a project, it is probably impossible to eliminate this risk completely.

The Risks

The impact of the unexpected discovery of contamination on a construction project can be great. For example, while excavating to install underground utilities, a public housing authority discovered that the project site had been used as a dumping site for industrial wastes. Removal of these hazardous wastes and cleanup of the site consumed the entire budget set aside for construction of the housing complex that was to occupy the contaminated site. The discovery of contaminants effectively doubled the cost of the project.

On another project, a contractor discovered chromium wastes while excavating the footings for a bridge pier. Though the contractor was not responsible for the costs, cleaning up the site resulted in substantial and costly delays to the project. The contract denied the contractor reimbursement for these delay costs.

These two examples illustrate the many problems that the discovery of contamination can cause on a construction site. Not only can such a discovery cause serious cost overruns, it can also lead to extensive project delays. The consequence of these cost overruns and delays may not be borne exclusively by the owner; as illustrated by the second example, the contractor may be required to bear some of these costs as well. There are also ways a contractor could become responsible for the cost of the cleanup itself. A key law addressing the responsibility for the cost of cleanup is the Comprehensive Environmental Response, Compensation, and Liability Act (CERCLA). This act provides that, in addition to current and former owners, parties that arranged for the disposal, treatment, or transport of the hazardous materials found on the site may also be responsible for cleanup costs. The terms *disposal* and *treatment* are broadly defined and include discharging, spilling, leaking, or otherwise depositing hazardous materials on land or in the water. It would be relatively easy for a contractor to transport hazardous materials unknowingly and deposit them elsewhere. For example, it is common for materials excavated from one site to be used as fill on another. If the excavated material was contaminated, the contractor could end up with responsibility for the cost of cleanup because it transported and deposited contaminated fill materials, even if the contractor was unaware that the materials were contaminated.

Contractors should also be aware that the discovery of contamination may jeopardize the financial health of the project owner or developer. Thus, even though the contractor may not be directly responsible for the costs of cleanup on a project where contamination is discovered, the contractor may find it difficult to obtain payment from an owner in financial distress.

Lenders are also concerned about their potential liability. This concern has two sources. The first source of concern is that the financial stability of a borrower may be harmed by the discovery of contamination on a site. The second concern is that the lender might become responsible for the costs of cleanup on foreclosed properties. In effect, the lender may become the owner. Under CERCLA, owners may be held responsible for the cost of cleanup.

With regard to the liability for the cost of cleanup, lenders have two potential defenses. In very particular instances, regulations provide that a lender's liability may be limited. With regard to underground storage tanks, for example, lenders meeting the defined criteria are not liable for cleanup costs under CERCLA. The applicable EPA regulation can be found in App. A.

The second defense provided for by CERCLA is known as the *innocent purchaser* defense. The term *innocent* in this case does not mean naive. The purchaser may not argue that it was simply unaware of the responsibilities it took on when it purchased the property or was unaware that the property purchased might be contaminated. Rather, an innocent purchaser must have used due diligence and carefully evaluated the site to determine whether contamination was present. If, after conducting such an evaluation and taking ownership of the property, it is discovered that contamination is present, the purchaser may be able to use the innocent purchaser defense to avoid CERCLA responsibility for cleanup costs.

For a lender to be able to use an innocent purchaser defense, it will typically require the borrower to evaluate the property as a condition of the loan. This evaluation is sometimes known as a *Phase I Inspection* or *Survey*. If such an evaluation is performed and no contamination is identified, then, should contamination be discovered later, the lender may be able to assert an innocent purchaser defense. Of course, knowing all the facts relevant to a particular situation is essential to any evaluation of liability. In addition, there is no substitute for qualified legal counsel in these matters.

In summary, the risks associated with the discovery of contamination are not limited to the project's owner or developer. Such a discovery also poses a significant risk to others, both on and off the project site.

Though iron-clad protections from the risks associated with the discovery of hazardous materials may not be available, there are several things that those in the construction industry can do to reduce their risks. The first, and perhaps most effective, is to perform an environmental audit. An environmental audit is conducted to evaluate a site's conformance with environmental laws and regulations. Other actions that can be taken include inserting appropriate protective clauses in real estate contracts, indemnity clauses, and insurance. Though contracts and insurance policies can be used to shift responsibility for the costs associated with the discovery of hazardous materials to others, these documents cannot prevent an owner or others from becoming involved with a contaminated project site in the first place. Avoidance of a contaminated site can only be accomplished through a careful evaluation of the site.

Evaluating a project site fully may entail a number of different tasks. Some of these tasks will be part of a normal environmental evaluation; some will not. The extent of the evaluation efforts will be a function of the perceived risks. For example, a commercial building built in the 1950s which has not had asbestos abatement performed will merit close inspection to determine the extent of asbestos contamination. A commercial building built in the 1980s will not need similar scrutiny. Given the potential variability, the discussion that follows is not a recipe for conducting an environmental audit, but a general approach to the evaluation of a potentially contaminated construction site.

The activities involved in the evaluation of a potential project site can be grouped into four categories: research, inspection, testing, and interpretation.

Research

The history of a building or site is likely to provide some indication of the probability that contaminants are present. For example, a site that previously housed a dry cleaner may very well be contaminated with dry-cleaning chemicals. The soils beneath former golf courses may be contaminated with arsenic. Older buildings may contain asbestos. Researching the history of a building or site may be an easy process or quite daunting.

There are several places to look. The first and, with luck, the best source of information on the project site is the site's current owners. There are two types of information that the current owner might possess. The first source is documentary. Some of the documents the current owner might possess that might be of interest to an evaluation of the project site for contamination are listed in Fig. 2.1.

Construction Documents

1. Plans and specifications of existing or demolished structures on site
2. As-built drawings of existing or demolished structures on site
3. Site utility plans showing existing or previous conditions
4. The construction documents relating to any abatement, remediation, or other cleanup activities on the site
5. Surveys of the site

Historical Documents

1. Aerial or other photographic views showing the site prior to development, demolished structures, the extent of remodeling or renovation, or existing structures
2. Books, magazine, or newspaper articles describing the site, its tenants or owners, the use of the site over the years, and the dates of major construction renovations or additions
3. Publicity or advertising material that describe the products or activities that occurred on the site
4. The results of any historic or other surveys or reports on the property, particularly information concerning the dates of development, and major renovations, prior owners or tenants, and the uses of the site

Reports and Evaluations

1. Geotechnical reports and evaluations
2. Records of borings, including, if retained, the borings themselves, boring logs, and evaluations
3. The results of any previous environmental audits, evaluations, or Phase I inspections or surveys
4. Title reports or other records of previous ownership

Figure 2.1 List of documents in possession of current owner of prospective site.

Oral interviews of the current owners of the property or facilities being considered for purchase may yield a significant amount of information as well. The term *current owner* includes the owner, the owner's agents, the owner's employees, tenants, and family members. The kind of information being sought includes a history of prior ownership or of prior tenants, the uses of the property, and the history of development on the site. Development includes initial construction, remodeling, renovations, additions, and demolitions. If materials were dumped on the site, the location of likely dump sites and a description of the kind of materials dumped will also be useful. A description of previous abatement, remediation, or other cleanup activities will be helpful, including a discussion of the purpose and scope of these activities. Finally, oral interviewers should find out if there are any others who should be consulted or questioned or if any additional documentation are available.

The current owner is not the only source of information. This is important to note if the current owner is not knowledgeable about the site or not forthcoming with information. Another good source of basic information about the site is the local government department or departments responsible for retaining the records of planning, development, construction, registration of deeds, and tax records. In larger cities or towns, some or all of these records may be maintained by the municipality itself. For smaller communities and rural areas, these records are generally maintained by the county. Wherever they are maintained, government records may provide the following basic information:

- Records of ownership
- Records of subdivision prior to development, and the associated local review and approval process
- Records concerning applications for building permits, escrow accounts maintained for public improvements, and the results of inspections
- Current and historical maps, models, and aerial photographs maintained by government agencies, which may show the existence of previous uses of or structures on the property. They may also allow the approximation of the date of development, construction, major renovations, or additions.
- Local histories describing the prior owners or uses of the property

Naturally, the quality of the information available from local governments is a function of their policies for retaining records. Typically, the earlier the date of the information sought, the more incomplete the records may be.

Beyond current ownership and local governments, potential sources of historical documentation concerning a site may be limited. The following are some possibilities:

- Prior owners or their employees or family members
- Libraries that maintain local historical records or information
- Contractors, suppliers, or other vendors that provided services to current or former tenants or owners of the property
- Users of the property
- Newspaper accounts of activities on the site, developments, construction, major renovations, and additions or important tenants or owners

Sometimes, multiple information sources can be used to piece together important events in a property's history. If local government records indicate the dates of major changes to or developments on the property, but do not describe the details of the development, newspaper accounts of the time may be able to fill in the missing details. The level of detective work necessary is usually a function of the site, its history, the potential for contamination, and the availability of information.

Inspection

Research may be insufficient to evaluate potential site contamination. The reasons may include the following.

1. Historical records are often limited or incomplete.
2. Memories are often hazy or wrong.

3. There is no guarantee that the site's current owners will be forthcoming with information, particularly if the information provided might affect the value of the property they are trying to sell.
4. Current owners, particularly banks, executors, trustees, realtors, or others who may have no personal experience with the property, may not be knowledgeable about it.
5. While historical data may indicate the possible presence of contamination, often the actual presence and extent of the contamination can be confirmed only by inspection.
6. The presence of some forms of contamination may not be revealed by research alone.
7. Records may not reflect the competence or success of previous abatement, remediation, or other cleanup efforts.

Increasing the chances of a successful inspection depends on a proper description of the scope of the inspection, selection of a qualified inspector or inspection firm, and providing the inspector with the resources necessary to conduct as thorough an inspection as possible. Determining the scope of the inspection is essential to selecting an appropriately qualified inspector. For example, some inspectors specialize in home inspection. While such inspectors might be perfectly adequate to conduct a home inspection before one purchases a home, they would probably not be qualified to inspect a former manufacturing plant. Thus, determining the scope of the inspection required is essential to ensuring that an inspector with the appropriate expertise and knowledge is selected.

The first consideration when developing an inspector's scope of work is to describe the nature of the site and how it is to be used. For example, are existing structures to be demolished and removed, or will they be renovated? If the buildings are to be renovated, how extensive will the renovations be? Will heating, ventilating, and air conditioning (HVAC) systems be removed and replaced or simply upgraded or repaired? Will interior partitions or ceilings be removed or left intact? How extensively will the site be disturbed? How extensive will excavations be? If these questions cannot be answered, then several alternates may have to be considered. If the design professionals for the project are already on board, their services may be enlisted to provide assistance in developing the scope of the site inspection.

In developing the inspector's scope of work, it may also be advisable to consider what the inspector should be searching for. For example, if an old school building is being considered for conversion to office space, the inspector should have experience appropriate to identifying likely hazards, such as asbestos, or lead paint, and any other hazardous materials that might be expected given the site's location and history. The preferred inspector for this job would also have some experience in the conversion of such facilities to commercial uses and thus understand the extent to which the existing building conditions will have to be disturbed to complete the renovation. The qualifications and expertise required for a school renovation project are probably quite different from those required to evaluate an existing golf course facility that is to be converted to a residential development. In addition, some consideration should be given to other things the inspector may be asked to evaluate during the inspection. Will the inspector be asked to inspect the existing buildings for structural integrity, for the condition of the roof, for the condition of the plumbing and electrical systems, for the efficiency and condition of the HVAC system, or the operation of other existing equipment or systems? Given the magnitude of the inspection needs, the services of more than one inspector or inspection firm may be required. Services beyond simple inspection may also be needed. For example, testing or estimating the costs of cleanup and other necessary work may be desired. All of these considerations will factor into the inspector's scope of work. It may be helpful to have a prospective inspection firm propose a scope of work as well.

In developing the inspector's scope of work, also consider the product required from the inspector. Indicate clearly whether or not a detailed written report is required. When evaluating this requirement, consider what might be needed by the lender, insurers, or others needing written documentation regarding the results of the inspection.

Once a detailed scope of work has been developed for the inspector, the selection process for the inspector can begin. Identifying potential inspectors or inspection firms might begin with referrals from a broker, an architect or engineer, a lender, or an attorney. In addition, inspectors or inspection firms might be members of professional or industry organizations. For example, home inspectors often belong to the American Society of Home Inspectors (ASHI). A list of member inspection firms for your area can usually be obtained by contacting these professional organizations.

Having obtained a list of prospective inspection firms, the next step is to decide which of the firms, if any, meets the needs of the project. The technical requirements are defined by the proposed scope of work. By combining these technical requirements with budgetary and timing requirements, the field can be narrowed before the final selection stage.

To be effective, the inspector must be provided with as much information as possible. Obviously, the inspector will need complete access to the site. Also provide the complete results of earlier research into the history of the property. This should include any as-built drawings, and other information that may have been obtained for the site and its structures. Discuss with the inspector things that are known or expected regarding potential contamination on the site.

The inspection will also have to be scheduled with the current owners. Scheduling should be done sufficiently in advance to allow the current owners to make any necessary arrangements so that the inspector can have full access to the site. The inspector may also have questions regarding the need for special equipment to gain access to certain areas of a structure. This might include items such as ladders, keys, or special operating devices or tools. If necessary, any security requirements must also be addressed.

Armed with the best information available, and given full access to the site, an inspector can help identify potential contamination if evidence of its presence is visible. An inspector cannot tell what might be behind walls, above ceilings, or under the floor or ground unless the inspector can gain access to these areas. In addition, an inspector will not be able to describe the extent of potential contamination. For example, though an inspector may be able to detect the presence of an abandoned and buried fuel tank, the results of the inspection may not show the contents of the tank or whether or not the tank leaked and contaminated the surrounding soils. Identifying the contents of the tank and determining whether or not the tank leaked must be determined based on tests.

Testing

Though an inspection of the property may indicate the potential for the presence of contamination, it may not be able to confirm that contaminants are present, indicate the extent of the contamination, or determine the concentration of the potential contaminants. If an inspector indicates that a potential exists for the presence of contamination, confirmation of the presence and extent of the contaminant may be required. When this information is needed, some testing may have to be done. If the inspection detected the presence of a buried tank and there is a need to know the contents of the tank, whether the tank has leaked or is leaking, and the extent of the contamination that may have resulted from such leakage, it may be necessary to conduct several tests. This testing might take the form of gaining access to and sampling the contents of the tank and taking borings around the tank to determine the extent, if any, of contamination resulting from leakage. Borings might also be required if the inspection suggests the possibility of buried wastes on the site.

How many borings are enough? What kinds of tests are needed? These are questions that neither your inspector, nor a testing firm may be able to answer. Ultimately, these may be questions only an expert in the cleanup of contamination can provide. This is particularly true if the contaminant has seeped into the groundwater supply. Evaluating the extent of such a problem can be quite costly and take a substantial amount of time. The cost of a cleanup may be even more substantial.

The process used to select a testing firm is similar to that of selecting an inspector. Selection of a testing firm should be preceded by development of a clear description of the testing work that is required. Based on references and research, a field of possible testing firms is identified. Through interviews and discussions this field is then narrowed to the testing firm that best meets the requirements of cost, timing, and qualifications.

Interpretation

Having performed extensive research, done a careful inspection, and performed all necessary testing, the last step is to interpret the data. Some interpretation work will be simple. If the inspector identifies the presence of materials containing asbestos, then asbestos is present on the site. If the potential purchaser's desire is to identify a site that is free of contamination, then this site is not what was sought. Other evaluations might be more difficult. For example, testing indicates that soils on the site contain lead. This information alone may not be sufficient to allow one to conclude that the site is contaminated. The next question to be answered by the evaluator is whether the concentration of lead in the soils on the site represents a hazard. If the concentration is low enough, the site may be considered to be clean or at least clean enough. Thus, in order to evaluate whether or not the site is contaminated to the extent that some remedial action is required, the evaluator must not only know that lead is present, but must know the concentration at which lead becomes a problem. For some contaminants, evaluation of the hazard represented by contamination based on inspection and test reports may be beyond the purchaser's capabilities, necessitating the hiring of qualified expert assistance.

In general, in the interpretation of the research, inspection, and testing data, the following questions may be asked.

1. Is contamination present on the site?
2. If contamination is present, does its location, concentration, or form make it a cause for concern?
3. Given the location, concentration, or form of the contamination, what remediation, abatement, or other cleanup activities will be necessary to develop the site as anticipated?
4. What is the estimated cost and estimated time to perform the recommended remediation, abatement, or other cleanup activities?

Consultant Selection

Clearly, someone or some firm capable of answering all of these questions for some types of contamination may possess more than ordinary experience or education. In fact, the qualifications necessary to answer these questions may be quite specialized. If a potential buyer of a property is seriously considering the purchase of a contaminated property, it is probably advisable to hire the kind of technical expertise required to answer the four questions listed. In fact, this level of expertise might be helpful throughout the project. For example, this expert could provide input into the project design phase, help develop a cleanup plan, and monitor the cleanup phase of work. This expert could also provide input to assure that all required permits, training, notification, and other actions are performed with regard to the hazardous materials encountered.

One avenue to selection of a consultant in the area of contamination is simply referral. Obtaining referrals and evaluating them can be a time-consuming process. Another approach to consider is used by many public agencies to select engineers, architects, and other design professionals and consultants. In this process, the public agency first develops a scope of work for the consultant. This scope of work is then advertised widely, and proposals are requested from interested parties. This advertisement is often called a *Request for Proposal* (RFP).

Proposals solicited from prospective consultants typically should contain the following information:

1. A detailed description of the consultant's approach to completing the scope of work.
2. A detailed description of the consultant's qualifications for performing the work, including the qualifications of key staff.
3. A listing and description of projects with which the consultant has been involved. The focus should be on projects similar to the project for which the consultant is being sought.
4. Other information may also be requested, but the first three items are essential. Other information may include availability and billing rates.

After reviewing the proposals submitted, a "short list" of three to five consultants is often prepared. The short-listed consultants are then given an opportunity to make oral presentations. This also provides an opportunity to ask specific questions about the consultant's background and experience.

Based on oral interviews, a final selection can then be made. Public agencies usually do not consider cost until after a selection is made. This is because cost may not be a good predictor of competence. Another approach is to consider costs only after qualified candidates have been short-listed.

Other Project Participants

This chapter has focused primarily on the needs of the prospective purchaser of the project site and the purchaser's lender. Other project participants, however, may need to evaluate the potential for encountering contamination before becoming involved with a project.

The contractor

The most concerned participants will probably be the contractors who will be working on the project site. Their concerns will be similar to those of the original purchasers of the site, but from a different perspective. Contractors will be concerned primarily with the risk associated with the project work. Contractors will likely have the following considerations.

1. The type, form, extent, and concentration of contamination that is to be encountered should be known.
2. If the presence of contamination is not indicated by the contract documents, the contractor may be interested in what contaminants are likely to be found on the site.
3. The contractor will be eager to know what its responsibilities will be with regard to remediation, abatement, or any other cleanup activities related to contamination.
4. Related to the cleanup work that is to be performed, the contractor will

be interested in the payment mechanisms associated with the performance of cleanup work.

5. The contractor will be interested in the contract provisions that relate to changes, differing site conditions, time extensions, and compensation for these items.

6. The contractor will be interested in the extent of exculpatory clauses, indemnification clauses, no-damage-for-delay clauses, and other similar contract clauses addressing the sharing of risk on the project. Of particular interest will be the site inspection clause. A *site inspection clause* is a contract clause that requires the contractor to inspect the site before offering a price to perform the work and enter into a contract. The clause generally attempts to make the contractor responsible for any site conditions that could have been detected by site inspection. The following is an example of a series of site inspection clauses from the standard contract provisions used by the New York State Department of Transportation.

> ARTICLE 3. EXAMINATION OF DOCUMENTS AND SITE. The Contractor agrees that before making his proposal he carefully examined the contract documents, together with the site of the proposed work, as well as its surrounding territory, and is fully informed regarding all of the conditions affecting the work to be done and labor and materials to be furnished for the completion of this contract, including the existence of poles, wires, pipes, ducts, conduits and other facilities and structures of municipal and other public service corporations on, over and under the site, and that his information was secured by personal investigation and research and not from the estimates or records of the Department, and that he will make no claim against the State by reason of estimates, tests or representations of any officer or agent of the State.
>
> 102.04 NO MISUNDERSTANDING. The attention of persons intending to make proposals is specifically called to "ARTICLE 3 of the AGREEMENT" wherein the bidder agrees that he has examined the contract documents and the site of the work and has fully informed himself from his personal examination of the same regarding the quantities, character, location and other conditions affecting the work to be performed, including the existence of poles, wires, pipes, ducts, conduits, and other facilities and structures of municipal and other public service corporations on, over or under the site, and that he will make no claim against the State by reliance upon any estimates, tests or other representations made by an officer or agent of the State with respect to the work to be performed under the contract. Particular attention is called to special notes and special specifications in the proposal which may contain contract requirements at variance with standard plans and specifications.
>
> The bidder's attention is also directed to the fact that in addition to his need to examine the contract documents and the site of work, there may be certain supplemental information which is available for his inspection in the Department of Transportation Office having jurisdiction for this project, as identified in the advertisement for bids. The supplemental information could include, for example, earthwork cross section sheets, various subsurface information, record plans, special reports and other pertinent project data. The proposal will include a list of the information available for inspection prior to the opening of bids.

102.05 SUBSURFACE INFORMATION. Boring logs and other subsurface information made available for the inspection of bidders were obtained with reasonable care and recorded in good faith by the Department.

The soil and rock descriptions shown are as determined by a visual inspection of the samples from the various explorations unless otherwise noted. The observed water levels and/or water conditions indicated thereon are as recorded at the time of the exploration. These levels and/or conditions may vary considerably, with time, according to the prevailing climate, rainfall and other factors.

The locations of utilities or other underground man-made features were ascertained with reasonable care and recorded in good faith from various sources, including the records of municipal and other public service corporations, and therefore the location of known utilities may only be approximate.

Subsurface information is made available to bidders in good faith so that they may be aware of the information utilized by the State for design and estimating purposes. By doing so, the State and the Contractor mutually agree and understand that the same is a voluntary act and not in compliance with any legal or moral obligation on the part of the Department. Furthermore, insofar as such disclosure is made, the Department makes no representations or warranties, express or implied, as to the completeness or accuracy of this information or data, nor is such disclosure intended as a substitute for personal investigations, interpretations, and judgment of the bidder.

Some owners may go so far as to require the prospective contractor to take responsibility for subsurface, concealed, or hidden site conditions based on the language in the site inspection clause. This is usually accomplished by offering the contractor the opportunity to conduct its own borings or excavate its own test pits to verify or evaluate subsurface conditions and then making the contractor responsible for subsurface conditions whether the contractor performs these site investigations or not. A site inspection or similar clause written in this form is a classic risk-shifting clause. It attempts to shift the risk associated with subsurface, hidden, or otherwise concealed site conditions to the contractor. Whether this clause can be used to make a contractor responsible for the unexpected discovery of hazardous materials and the associated remediation, abatement, or other cleanup work is debatable. Such a clause, however, would on its face dramatically increase the risk associated with a contractor's potential work.

7. In concert with the site inspection clause, the contractor will be eager to know what other descriptive information is available. This information might include boring cores, boring logs, geotechnical reports, the results of environmental audits and evaluations, as-built plans and records from previous work on the site, and any other documents that the contractor is provided that describe the existing site conditions.

It is generally in the owner's best interests to disclose all information in its possession concerning site conditions. The primary reason for this is that the owner would prefer that the contractor's price consider as many potential problems as possible. The owner should also be aware that failure to disclose material information may allow the contractor to skirt some of the risk-shifting provisions in the contract.

For example, consider a project where asbestos abatement is the contractor's responsibility. The contractor underestimates the quantity that must be removed and seeks compensation from the owner for the overrun. As it turns out, the owner was in possession of annotated as-built drawings showing the extent of asbestos contamination throughout the building. Had the contractor been provided with these annotated drawings, it could have properly estimated the extent of the contamination. In such a scenario, the owner might be responsible for the overrun, particularly if the contractor's original estimate was a good one given the available documents.

8. To evaluate its risks fully, the contractor may also have to learn about the applicable permitting, posting, training, site access, notification, and other regulatory considerations associated with working on a contaminated or potentially contaminated site.

All of these considerations will enter into the contractor's decision to pursue and accept work on the project and the price and contract terms offered to do the work.

The designers

The designer's concerns regarding potential contamination will center around the design considerations, needed expertise, and liability. The latter two factors will probably influence the designer's decision most in terms of whether to pursue work on a potentially contaminated project. The designer's chief risk is that it will negligently fail to consider the presence of contamination adequately in its design. If the designer's experience with the hazardous material and knowledge of the project site is such that it can confidently prepare an adequate design, then it can confidently pursue work on the project. If not, then it will either have to supplement its staff with the appropriate level of expertise or forgo the work. The designer may also want to persuade the owner to perform adequate site investigations to ensure that the potential hazards are evaluated adequately. This would include sufficient borings, test pits, and other samples and tests.

Summary

All parties to a construction project are interested in evaluating the risks associated with potential contamination. Each evaluates these risks using a combination of research, inspection, testing, and expert evaluation. The focus of each party, however, may be different. The owner is trying to assess whether the presence of contamination threatens the feasibility of the project intended for the site. The lender is trying to ensure that the risks and costs of contamination have been evaluated adequately. The contractor is trying to determine whether the potential profits merit the risks to be incurred. The designer is interested in ensuring that it has adequate expertise and knowledge of the project conditions to prepare a competent design.

Chapter 3

Preemptive Actions before the Project Begins

Despite the best efforts of those involved, or perhaps because those involved did not give their best efforts, contamination may still be discovered on a construction project. The risk of discovery is a risk that can be reduced and managed but not eliminated. As discussed in Chap. 2, the risk of discovery in the first place can be reduced by carefully researching the project site, inspecting the site, conducting appropriate tests, and properly interpreting the test and inspection data that are collected. Though the risk of discovery probably cannot be reduced further, the effect that such a discovery might have on a project can be reduced. This reduction can be achieved through the use of good project management practices, the subject of this chapter.

Project management consists of two basic activities, planning and execution. *Planning* consists of the steps taken before performing the work to assure that the work proceeds smoothly and to increase the chances that the project will be completed on time, within budget, with the expected profit, and with appropriate levels of quality. *Execution* consists of both actions taken to implement the plan and efforts expended to deal with problems that threaten the successful completion of the plan. The discovery of contamination falls into the category of problems that threaten the successful completion of the plan.

Planning

Good planning is essential to reducing the impact of the discovery of contamination on the project. Planning to reduce the impact, however, is not something that can be done after the contamination has been discovered: It must be done in advance of discovery. In fact, planning for the possibility of discovering contamination is something that should be done before any construction begins. This is because many of the management systems that are required to minimize the impact of such a discovery must be in place before construction begins. For example, a termination for convenience contract clause is essential to the orderly abandonment of a project that is rendered uneconomical by the discovery of contamination. Such a clause must be inserted into the contract between the owner and the contractor before it is signed. It is very difficult to negotiate the addition of this clause once the project has begun. Similarly, a contractor offering a fixed price to complete a project must factor in the cost of training its crews to recognize, report, and, perhaps, mitigate the spread of the types of contamination that are likely to be found on a project. It is too late to add these costs to the contract price after the price has been negotiated and agreed to by the owner.

Contamination threat assessment

The first step in planning a project where the discovery of contamination is a possibility is to identify the likely contaminants and likelihood of their discovery. This might be called a *contamination threat assessment*. Some purchasers have such an assessment performed before purchasing the site in the first place. If the site was evaluated as described in Chap. 2, then the purchaser is already aware of the contaminants, if any, that are likely to be discovered during construction. For other project participants, however, or in situations where the owner is renovating existing facilities or building on land that has never been assessed, evaluating the risk of discovering contamination is an essential first step to planning how to deal with such a discovery. This evaluation process is described in detail in Chap. 2.

Though the evaluation process may be similar, each party will be seeking different information. The owner will want to know chiefly what kinds of contaminants will most likely be discovered and how such a discovery will affect the cost and timing of development. The architect and engineer will be most concerned about how the project design can best address the likely hazards. The contractor will be most concerned about the kinds of hazards its crews might encounter and the party that will be responsible for the costs associated with the discovery, containment, and cleanup of any contaminants discovered.

Because the objectives and concerns differ, the planning efforts undertaken by each project participant will vary as well.

The owner's planning efforts

The first project planning efforts will almost always be undertaken by the owner or developer of the project. Most of these initial planning efforts will be devoted to determining what is to be built, site selection, budgeting, and financing. The owner will also have performed a contamination threat assessment and determined what contaminants are likely to be encountered, if any. After deciding what is to be built, selecting an appropriate site, developing a budget, lining up financing, and evaluating likely contamination risks, the owner's primary planning responsibilities are the selection of the project team, including the designer and the contractor or contractors; determination of the contractual relationships among the team members; determination of the risks to be borne by each party regarding the discovery of contamination; and project timing.

Selection of the project team. The makeup of the project team can vary significantly depending on the size and complexity of the project and the owner's in-house capabilities. Typically, however, the owner will require the services of a design professional, usually an architect or engineer, to design the project and a contractor to build the project. The design team may consist of a single firm or separate architects, engineers, and space and landscape designers. The contractor may also be a single company or multiple contractors managed by the owner or a construction manager. The owner might also hire a firm that performs both design and construction (a design/build firm). Whatever project team the owner develops, and the possible permutations are almost endless, the owner's chief concern with regard to the discovery of contamination is to select architects, engineers, and other team members that have adequate knowledge and experience with the contaminants that are most likely to be discovered. Thus, if it appears likely that asbestos is present in a building to be renovated, then it makes sense for the owner to consider the prospective contractor's experience with regard to asbestos.

Using experience with asbestos or some other contaminant as a selection criterion for picking the members of the project design team is not difficult. As discussed in Chap. 2, a typical process for selecting architects, engineers, and other design professionals is the Request for Proposal process. In this process, the owner advertises for proposals from those interested in performing the work. Advertisements may be to the industry as a whole, or a combination of advertisement and invitation. In addition to requesting general design qualifications, the RFP would ask proposers to describe their qualifications with regard to the contaminants of interest. The RFP might also ask the proposers to describe their basic approaches to handling these contaminants in their designs and in the construction documents. After narrowing down the field of responses to the RFP to a short list of qualified candidates, oral interviews are often conducted. During these oral interviews, the candidates might be asked probing questions about their experience with the contaminants of interest and about their approaches to the handling of these contaminants on the site. A sampling of the kinds of questions that respondents might be asked to address is provided in Fig. 3.1

Based on the designers' written proposals, oral presentations, responses to questions, availability, costs, and the comments made by references, the owner can make an informed decision about the selection of a designer.

The process by which the contractor is selected is often different from the process used to select the designer and other professional consulting assistance. In addition, the selection process for contractors may vary depending on whether the owner is from the public or private sector. In the private sector, the contractor is usually prequalified, then competes on the basis of price. In the public sector, the contractor is more likely to be selected purely on the basis of price.

In the private sector, the first step in the selection of the contractor is often to identify a pool of qualified contractors. This pool is identified based on recommendations from the designer and others in a position to recommend qual-

1. What experience has the firm had with the types of hazardous materials that are present or are most likely to be found on the project?
2. With regard to this experience, what specific projects has the firm been involved with?
3. What was the firm's role on each project?
4. Did the firm hire outside consulting assistance to handle this work?
5. Are the employees that lead the firm's efforts on these projects still with the firm?
6. What are the names and telephone numbers of references that could be contacted regarding the firm's performance on these projects?
7. What are the qualifications and experience of the persons proposed for work on the project?
8. How soon and for how long will the proposed project team be available?
9. On what other projects are the various project team members working?
10. With regard to the firm's approach to addressing the contamination issues on the project, has the firm used this approach successfully before?
11. Where was the approach used by the firm before?
12. What are the names and telephone numbers of references that can be contacted regarding the firm's proposed approach to handling the hazardous material of interest?
13. What problems does the firm expect to encounter with regard to the design and application of the firm's approach to handling the hazardous material of interest?
14. What research or other information that is not already available must be acquired in order to prepare the firm's design?
15. How long will the firm need to perform its design work?
16. How will the firm manage this work, including how does the firm schedule and budget its work?

Figure 3.1 Interview questions for selection of design professionals.

ified candidates. The private owner may also advertise for proposals. From this pool of potential contractors, the field is often narrowed to a group of three to five qualified candidates. These contractors are then provided with a set of bid documents and asked to provide a price for performing the work. The owner then selects the contractor that provides the best combination of price, qualifications, reputation, availability, and terms. When selecting the contractor, the owner may consider the contractor's qualifications with regard to the hazardous materials that are present or likely to be found on the site. The owner may also seek the contractor's answers to the questions listed in Fig. 3.1, modified as appropriate for interviewing a contractor.

In the public sector, the prequalification and selection process just described may prove to be problematic. It is often difficult for the owner to reduce the potential field of competitors through prequalification. It is not impossible, however, to prequalify on public projects, particularly if the basis for the prequalification requirements are objective and measurable. The following contract language was used to prequalify bidders on a bridge project.

> 102.01 Qualification of Bidders. All bidders must be prequalified by the Department to submit a proposal. A bidder seeking prequalification shall complete the "Preaward Qualification Questionnaire for Bidders." The executed form shall be that which is contained in the bid package and shall be completely filled out and submitted with the proposal.

Attendance at the mandatory pre-bid conference is a prerequisite for bidding for all contracts.

A "Special Contractor Prequalification Questionnaire" for bidders, used for this project, is provided in App. B.

Contractors that successfully complete the prequalification questionnaire and meet the basic prequalification requirements are then allowed to submit a bid on the project. The prequalification process used on the bridge project was not challenged by prospective bidders that were not qualified to bid the project.

The process used by the owner to select the construction manager is similar to the process used to select the designer.

Contract relationship. The owner is typically in control of the contract relationships among the project team members. For example, the owner decides whether or not the work is contracted to a single general contractor or to multiple prime contractors. The owner determines whether or not a design/build team is appropriate for the project. These decisions are not usually made by other project team members. The types of relationships established on projects are more a function of the experience of the owner and the requirements of the project than the likelihood that contamination might be discovered. In other words, no contractual relationship is inherently superior to another when the discovery of contamination is a possibility. The potential for the discovery of contamination, however, does increase the complexity and level of sophistication required of the project team, including the owner. At some point, the requirements of the project may exceed the owner's capabilities. In such a situation, the owner will need to bring aboard assistance to expand the owner's capabilities and act as an extension of the owner's staff or as the owner's representative. This party is often known as the *construction manager* (CM), *project manager* (PM), or *project coordinator* (PC). There are many names by which a firm that assists an owner may be labeled. The function, not the name, is significant. For the remainder of this discussion, the owner's assistant will be referred to as the construction manager.

To a large extent, the role the CM plays on a project and the effect this role has on the contractual relationships among the parties on the project is a function of how the CM is paid. Some CMs act primarily as the owner's representative, filling the roles traditionally filled by the owner's staff, plus taking on the project administration tasks that are often performed by the project architect or engineer. CMs acting in this capacity are typically compensated on a cost-plus basis. In other words, they are reimbursed for the actual costs of performing the work, plus some markup to cover overhead and provide a profit. This CM role is sometimes referred to as an *agency CM*. Some CMs work on an "at-risk" basis. In situations where the CM is being compensated on an at-risk basis, the CM often agrees to complete the project work for a *guaranteed maximum price* (GMP). The CM is compensated for all costs incurred to construct the project, plus some markup for overhead and profit, up to the GMP. If the cost of constructing the project exceeds the GMP, then the CM becomes responsible for these added costs. That is why CMs compensated in this manner are considered to be "at risk." There are arguments for and against either approach to paying the CM. For example, some might argue that because the CM at risk is responsible for the overall cost of the project, it will be more concerned about controlling overall project costs. Alternatively, because CMs working on an agency basis receive all of their compensation directly from the owner, and have no contractual allegiances to other parties on the site, they may be perceived as acting solely as the owner's representative and thus more truly provide loyalty and better represent the owner's interests. Therefore, in situations where the owner is hiring a CM to help bolster its own forces, an agency CM would appear to be the logical choice.

Typically, on projects with an agency CM, all the other major parties on the project contract directly with the owner. In fact, in some instances, even major subcontractors and suppliers contract directly with the owner. In these cases, the CM is charged with the overall responsibility to coordinate, administer, and manage these contracts. The roles and responsibilities of each of the parties to such a contract are set forth in their respective contracts. Because these contracts must be developed before the parties are brought on board, they are part of the early project planning efforts that are the responsibility of the owner.

Figures 3.2 through 3.5 show several approaches to the basic organization of the project team.

The traditional project team. Figure 3.2 depicts a traditional project organizational structure. The designer is contracted directly to the owner. In addition, the designer provides contract administration services. To provide these

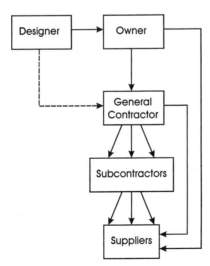

Figure 3.2 Traditional approach to the organization of the project team.

administrative services, the designer may deal directly with the contractor, though there is no contractual relationship.

In the traditional approach, a general contractor also contracts directly with the owner. The general contractor subcontracts work that it cannot perform economically. Suppliers in turn are sub-subcontracted by subcontractors to provide materials and equipment to the project. Both the owner and the general contractor may also have contracts with suppliers to provide materials and equipment to the project site.

The construction manager at risk. When a CM is used and this CM is at risk, the organization of the project team is very similar to the traditional organization (see Fig. 3.3). The only difference is that the designer's contract administration role is often substantially reduced. The at-risk CM holds subcontracts with subcontractors. The suppliers may still be sub-subcontracted. Both the owner and the at-risk CM may maintain contract relationships for the supply of equipment and material for the project as well.

The agency construction manager. Figure 3.4 depicts a typical contracting arrangement where an agency CM is used. The owner holds contracts with the designer, the agency CM, multiple prime contractors or a general contractor with subcontractors, and, perhaps, even some suppliers of material and equipment. All contracts held by the owner, however, are managed and administered by the agency CM.

The design/build team. Diagrammatically, the simplest contract organization is represented by the design/build approach to construction. In this approach,

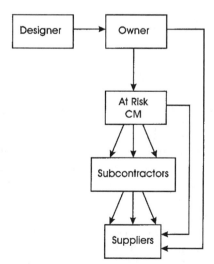

Figure 3.3 Construction-manager-at-risk approach to organization of the project team.

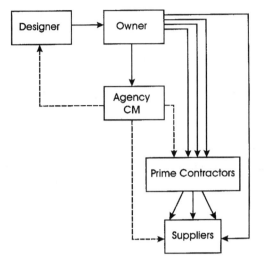

Figure 3.4 Agency CM approach to organization of the project team.

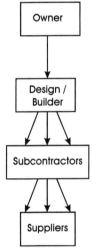

Figure 3.5 Design/build team approach to organization of the project team.

depicted in Fig. 3.5, the designer and contractor are one entity. Subcontractors and suppliers are subcontracted to this entity.

Assigning risk. The contract is the primary tool for risk assignment on a construction project. The owner is usually, though not always, responsible for developing the contract documents, particularly the risk management clauses in the contract general provisions. For this reason, developing the contract is an early responsibility of the owner.

The risks associated with the discovery of contamination are addressed by a number of clauses in the general provisions of the typical construction contract. In addition, the contract is often used to mandate some of the management tools used to reduce the impact of the discovery of contamination on a project. For example, contracts often contain detailed scheduling clauses that not only describe the type of schedule that must be used, but how this schedule will be used to manage the project. Typical contract clauses and their purposes are discussed in Chap. 4.

Owners should be careful when using clauses which attempt to shift risk. For example, a common risk-shifting contract clause is a "no-damage-for-delay" clause. Such clauses are also known as *exculpatory clauses*, because they attempt to exculpate or excuse the owner from some risk. A no-damage-for-delay clause states that in the event that the project is delayed, the contractor will not be entitled to reimbursement of the costs associated with the delay. Such a clause makes good sense when the delay was the contractor's fault or could have been anticipated. It becomes a risk-shifting clause when it is extended to include delays that were the fault or responsibility of the owner. Its application to the discovery of contamination is that while the contractor might be entitled to recover the extra costs associated with the discovery, it would not be entitled to recovery of the costs associated with project delays resulting from the discovery of contamination.

When using no-damage-for-delay and other exculpatory contract clauses, owners should be aware of the possible consequences. Exculpatory clauses are contract clauses that free a particular contracting party from responsibility for certain problems. For example, a no-damage-for-delay clause frees an owner from responsibility for the costs associated with project delays, whether or not these delays were caused by the owner. The following is a typical no-damage-for-delay contract clause.

> ARTICLE 13. DELAYS, INEFFICIENCIES, AND INTERFERENCE. The Contractor agrees to make no claim for extra or additional costs attributable to any delays, inefficiencies, or interferences in the performance of this contract occasioned by any act or omission to act by the State or any of its representatives. The Contractor also agrees that any such delay, inefficiency, or interferences shall be compensated for solely by an extension of time to complete the performance of the work in accordance with the provisions of Subsection 108-04 in the Standard Specifications. In the event the Contractor completes the work prior to the contract completion date set forth in the proposal, the Contractor hereby agrees to make no claim for extra costs due to delays, interferences, or inefficiencies in the performance of the work.
>
> The Contractor further agrees that he has included in his bid prices for the various items of the contract any additional costs for delays, inefficiencies, or interferences affecting the performance or scheduling of contract work caused by or attributable to, the following instances:
>
> 1. The work or the presence on the contract site of any third party, including but not limited to that of other contractors or personnel employed by the State, or by other public bodies, by railroad, transportation or utility companies or corporations, or by private enterprises, or any delay in progressing such work by any third party.

2. The existence of any facility or appurtenance owned, operated, or maintained by any third party.
3. The act, or failure to act, of any other public or governmental body, including, but not limited to, approvals, permits, restrictions, regulations or ordinances.
4. Restraining orders, injunctions, or judgements issued by a court.
5. Any labor boycott, strike, picketing or similar situation.
6. Any shortages of supplies or materials required by the contract work.
7. Climatic conditions, storms, floods, droughts, tidal waves, fires, hurricanes, earthquakes, landslides, or other catastrophes.
8. Determination by the Department to open certain sections of the project to traffic before completion of the entire contract work.
9. Increases in contract quantities, additional contract work, or extra work or delays in the review or issuance of orders-on-contract, or shop drawings, or field change sheets.
10. Failure of the State to provide individual rights-of-way parcels for an extended period of time beyond that indicated by the contract.
11. Unforseen or unanticipated surface and subsurface conditions.
12. Stop work orders issued by the Engineer, pursuant to Subsection 105-01.
13. Any situation which was, or should have been, within the contemplation of the parties at the time of entering into the contract.

One of the results of using exculpatory provisions to shift risk to other project participants is that these parties then perceive the project to be more risky. A riskier project may mean a higher price. A riskier contract may also mean that fewer contractors will be interested in performing the work. Reducing the number of contractors interested in competing for the work may reduce competition and result in higher bids. Because exculpatory clauses may be perceived to be basically unfair, such clauses may also make it more difficult for the project participants to come together and form a cohesive team. This lack of team spirit can make disputes more difficult to resolve and may increase the likelihood of claims being filed and litigation ensuing during the project. Even worse, in exchange for a higher price and the increased likelihood of disputes, the owner may not be getting the kind of protections expected from the exculpatory clause. For example, in California the inclusion of a no-damage-for-delay clause in some contracts is unenforceable. In many legal jurisdictions, courts and government appeals boards have sometimes refused to uphold exculpatory clauses. Standard exceptions to enforcement of these clauses have been recognized. For example, with regard to no-damage-for-delay clauses, in situations where the delay resulted from a contract breach, there was active interference by the owner, there was evidence of fraud, or the delay could not have been anticipated or was of an unusually long duration, some judicial bodies have failed to enforce the clause. Given the difficulties of enforcement, the potential to increase the cost of construction, and the increased chances of claims and litigation, the use of exculpatory clauses is not advised.

Risk is also a function of the form of the contract. There are essentially two basic types of construction contract forms. The first basic type is a *fixed-price* or *lump-sum contract form*. This is the most common type of contract between an owner and a contractor and between a general contractor and its subcontractors or suppliers. In a lump-sum contract, the contractor is paid the lump-sum amount of the contract for completing the contract work. If the cost of completing the contract work is less than the lump-sum amount, the contractor pockets the difference as profit. If the cost exceeds the contract amount, the contractor's compensation is capped at the contract amount and the contractor suffers a loss. This kind of contract works best when the work required by the contract can be clearly and completely defined. It shifts much of the risk associated with cost overruns on the contract work to the contractor.

The second basic contract type is called a *cost-plus* or *time-and-materials contract*. In this type of contract, the contractor is reimbursed the direct cost of the work, plus some percentage markup to cover overhead costs and profit. In this kind of contract, the incentive for the contractor to keep costs to a minimum is weaker. In addition, the owner now bears the risk that the cost to complete the contract work may exceed the estimated cost, but the owner will benefit if the work is completed for less than planned. Cost-plus contracts are common in situations where the scope of the contract work is difficult to define. It is most commonly used in contracts with designers, CMs, and other professionals, and for some types of extra or emergency work.

There are many variations on these two basic contract formats. One common variant of a fixed-price contract is a unit-price contract. In a *unit-price contract* the contractor provides prices to perform specific units of work. For example, a contractor might provide a price to pour a cubic yard of concrete. The price would include the cost of constructing the formwork, supplying the concrete, the cost of placing, finishing, and curing the concrete, and other incidental costs. The unit price would also include the cost of stripping the forms. The unit price, however, would apply only to a single cubic yard of concrete. The contract price of a unit-price contract consists of estimated quantities provided by the owner, to which the contractor applies a unit price. The product of these two numbers is the estimated cost of a particular contract item. When added to the other contract items, a total price to construct the project results. In a unit-price contract, the contractor bears the risk that the contract items cannot be completed for the unit price bid. The owner, however, bears the risk for quantities that are underestimated. Unit-price contracts are appropriate when the costs to perform a particular unit of work are well known, but the quantity is not known with precision. Highway projects are often built using unit-price contracts. Hazardous-waste cleanup projects are also often completed under a unit-price contract.

Another variant of a lump-sum project is a lump-sum project with certain items provided for on a unit-price basis. These unit-price items in a lump-sum contract are sometimes called "if and where directed" items. The following is an example of an "if and where directed" item contract clause.

"If and Where Directed" Items

The Proposal Form may request bids on one or more Pay Items to be incorporated into the Project "if and where directed" by the Engineer. The Engineer shall have sole discretion in determining whether and to what extent such items will be incorporated into the Project. Incorporation of such items into the Project shall only be made on written direction of the Engineer. In the absence of written direction, no such items shall be incorporated into the Project and if incorporated shall not be paid for. The Engineer may order incorporation of such items at any location within the contract and at any time during the work. These items will not be located on the Plans. The estimated quantities set out in the Proposal for such items are presented solely for the purpose of obtaining a representative bid price. The actual quantities employed may be only a fraction of, or many times the estimated quantity. The Contractor shall make no claim for additional compensation because of any increase, decrease, or elimination of such items.

The use of this type of contract clause is particularly applicable to a project where the discovery of contamination is possible. For example, for projects where asbestos is likely to be encountered, a unit price for asbestos removal may be provided in the contract, together with an estimated quantity. This allows the owner to establish a competitively bid price for the work in the contract, but does not assign to the contractor the risk of having to remove an unspecified quantity of asbestos. Given the difficulty of finding and quantifying the amount of asbestos that might have to be removed, this is a reasonable way to share the risks involved.

Pure cost-plus contracts are often modified to provide some incentive to the contractor to control costs. One way to do this is to fix the fee. A *fixed fee* is a designated fixed amount set aside as the profit or fee the contractor is to receive for completing the contract work. The fee is not a percentage of costs, but a fixed amount. In this arrangement, the more the contractor spends and is reimbursed to perform the work, the smaller its profit is as a percentage of its overall payment. Controlling costs will allow the contractor to realize a higher profit percentage.

A *not-to-exceed price,* sometimes called a *guaranteed maximum price* (GMP), might also be established on a cost-plus job. This establishes a cap on the owner's costs (assuming no changes) and encourages the contractor to control costs.

A third technique to provide an incentive to control costs on a cost-plus job is used in conjunction with a GMP. If the contractor is able to complete the work for less than the GMP, the owner and the contractor share in the savings. This establishes a reward to the contractor for successfully controlling costs. It is not uncommon on a cost-plus job for all three of these techniques to be used. A GMP may be established, with a fixed fee and sharing in the savings.

Different kinds of project team organizations lend themselves to different kinds of basic contract types. Figures 3.6 through 3.9 identify the kinds of contract types that are found among the various parties in each of the four project team organizations. Typically, contracts between designers and owners are cost-plus contracts. Fixed-fee and GMP contracts may be used to help

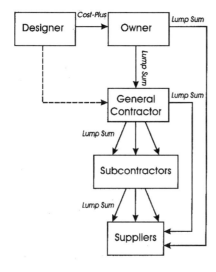

Figure 3.6 Contract types for a traditional project team approach.

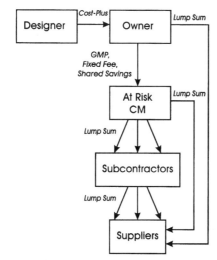

Figure 3.7 Contract types for a project team with a construction manager at risk.

incentivize cost control. Lump-sum contracts are less common. Between owners and contractors, lump-sum contracts are the most common. The same is true of contracts among contractors and subcontractors and suppliers.

Contracts between an owner and an at-risk CM are often cost-plus contracts and almost always include a GMP. Fixed fees and a mechanism to share in the savings are also often included. Contracts between an at-risk CM and a subcontractor or supplier, however, are often lump-sum contracts. Between an agency CM and an owner, the contract type used is usually cost-plus. A fixed fee is often used in this case to provide cost control incentive for the CM.

For design/build projects, the lump-sum contract format is most common among the owner, contractor, subcontractor, and supplier.

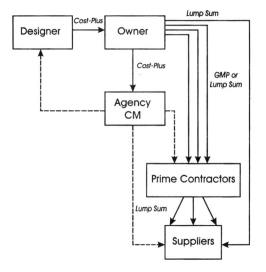

Figure 3.8 Contract types for a project team with an agency construction manager.

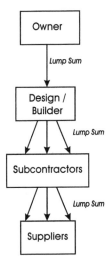

Figure 3.9 Contract types for a project with a design/build team.

Project timing. The typical approach to timing the major events of a project is first to prepare the project design and then to select a contractor and perform the work. Major events are sequential—first design, then build. But what if there is not enough time to build a project sequentially? Must the design be 100 percent complete before any construction can take place? The truth is that the sequential execution of major project events is not necessary. By "fast-tracking" a project, construction can begin before the design is completed. The idea of fast-tracking is that construction can begin on certain items as soon as the design of these items is completed. There is no need to wait for the entire design to be completed. For example, construction can begin on a building's foundations before the specification of the building's interior finishes is completed. As long as the design for each part of the project is completed so as not to hold up construction, design and construction of the project can proceed concurrently or in a phased fashion. The concept of fast-tracking is potentially quite useful on a project where the presence of contamination is known. Using a fast-track approach, remediation contractors skilled in the cleanup of a particular contaminant can be mobilized on site before construction of the larger project is scheduled to begin. For example, an asbestos-removal contractor could be brought on site to clean up the asbestos before renovation of the contaminated facility begins, or even before the design of some portions of the project are totally complete. In most cases, fast-tracking will be the preferred approach to the timing of projects where remediation, abatement, or other cleanup activities are required.

Owners should keep in mind, however, that a fast-tracked project is significantly more complex to manage and administer. This added complexity comes from the fact that design and construction activities are proceeding concurrently. It is also more complex because it requires a much tighter meshing of the design and construction processes. In fact, it is essential on a fast-tracked project that the design and construction phases be scheduled together on a single master project schedule. Figure 3.10 presents a simplified barchart depiction of a fast-tracked project where asbestos abatement is required before the start of other construction activities.

Because fast-tracked projects are more demanding of project management and administration time, owners often recognize a need for assistance to manage and administer fast-tracked projects. This assistance is often brought

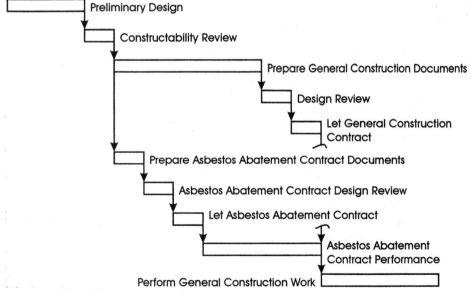

Figure 3.10 Fast-tracked schedule for asbestos abatement.

on at the project's inception. Because the term "construction manager" does not describe the full range of services that are provided, this type of assistance is often provided by a program manager (PM). A PM's role may include the following activities:

1. Management and administration of feasibility studies
2. Management and administration of borings, excavation of test pits, and other tests
3. Preparation of contract documents for and assistance in the selection of design consultants
4. Management and administration of design consultant contracts
5. Development of overall program schedule
6. Development of individual bid packages for each phase contractor
7. Assist in the selection of contractors
8. Management and administration of construction contracts
9. Design reviews
10. Constructability reviews
11. Development of overall program budget
12. Review of invoices and pay applications
13. Review of requests for contract modifications
14. Negotiation of contract modifications
15. Monitoring of progress
16. Inspection services
17. Facility testing and start-up activities
18. Estimating services
19. Document control
20. Organization, scheduling, and preparation of minutes for meetings
21. Preparation of daily reports and logs
22. Preparation of monthly project status reports
23. Punchlist inspections
24. Review and evaluation of claims

A PM's contract is usually cost-plus, often with a fixed fee.

In summary, with regard to planning for the possibility that contamination might be discovered on the project, the owner bears the most significant responsibilities, and these responsibilities must be met earlier than those of any other project participant. In addition to deciding what is to be built, and site selection, the owner must choose project team members who have sufficient knowledge and experience to deal with likely contaminants. The owner must decide how the team members will relate to each other contractually and assure that, if its own forces are lacking, adequate management assistance is brought on board. In addition to deciding how the project parties are to relate contractually, the owner often has to determine the contract forms that are appropriate for the project and the appropriate contract clauses. In considering what clauses are to be included in the contract, the owner must decide how risks are to be shared by the contracting parties, being careful to recognize that one cannot necessarily eliminate risk by attempting to place it on another party. The owner can manage and reduce risk, but cannot eliminate it. The owner is also ultimately responsible for determining the sequencing and timing of the major contract activities and deciding whether fast-tracking is necessary and appropriate.

The designer's planning efforts

Chronologically, the designer's planning efforts are performed after the owner's. With regard to contamination, the designer's planning efforts will depend on the extent of the designer's responsibilities on the project. For example, if the designer is charged with preparing all contract documents for the owner, then the designer will have to consider some of the same planning questions assigned to the owner. Will the work go best with a fast-track or phased approach? Will contracts be lump-sum or cost-plus, or some hybrid of the two? What types of clauses should be used to manage risk?

The designer's chief planning responsibility, however, will be the design of the project. With regard to possible contamination, the designer will have two primary responsibilities. The first is to become as familiar as possible with the potential contamination risks on the project. Since the designer's experience with certain types of contamination may have been one of the criteria in the designer's selection in the first place, the designer may already have some basic familiarity with the hazardous materials of interest. In becoming familiar with the project, the designer will be trying to identify the likely contaminant. The designer will also be attempting to learn the extent of the problem, including the probable locations and likely amounts of contamination. Finally, the designer must become familiar, if he is not already, with the means by which the contamination is to be cleaned up or contained. This may include becoming familiar with remediation techniques, and researching permit and other regulatory requirements. Research approaches to acquiring the necessary information are discussed in Chap. 2.

As part of the research effort the designer should quickly identify whether additional testing or expertise or both are required to evaluate the type, location, and extent of contamination, and its cleanup. If additional testing or help is required, it is important for the designer to notify the owner as quickly as possible, so that this additional testing and analysis work does not hold up the progress of the design or the start of construction. Keep in mind that a day of delay in the design process may have the same effect as a day of delay during construction. Either delay may postpone the ultimate project completion date or increase the cost of the project. Given the importance of time, it is generally a good idea for the designer's work to be scheduled tightly.

The objective of learning all there is to know about the contaminant or contaminants of interest is to allow the designer to develop a design for the project that deals with the contaminant as required by applicable regulations in the most economical way possible given the requirements of the project. For example, lead-based paint was used for decades on and in buildings throughout the country. When considering the renovation of a structure, the architect will want to know whether lead-based paints are present. If lead-based paint is present, the architect will want to know whether or not the paint must be removed, given the intended uses of the rooms to be renovated. If removal is required, the architect may have to develop specifications for the paint's removal. From a broader perspective, the design professional may be charged with considering alternatives. With regard to lead-based paint, for example, it may be preferable to cover contaminated areas rather than remove the paint. The designer may also be charged with what amounts to feasibility studies: weighing the advantages and disadvantages of various approaches to the problem.

For example, given the extent of the contamination and the nature of the necessary renovations, might the most economical solution be demolition and new construction rather than renovation? Alternatively, the designer may have to evaluate different renovation schemes that might reduce or eliminate the need for removal of the hazardous material.

The contractor's planning efforts

Unless the contractor is brought onto the project very early, planning efforts with regard to the possibility or presence of contamination on the site will focus primarily on the costs and risks. The contractor's planning efforts will also involve a substantial amount of research. The objective of this research will be to allow the contractor to develop a cost for dealing with the contamination. While performing this research, the contractor will consider the requirements of the plans and specifications. Other than obvious (patent) errors or other problems, the contractor will have to assume that the plans and specifications represent accurately the requirements that the contractor must fulfill in order to complete the work. If obvious problems are identified, the contractor generally has an obligation to make the owner aware of these problems. The contractor should not be expected to catch hidden (latent) problems within the plans and specifications. In addition, it is unreasonable to expect a contractor to duplicate the expertise of design professionals. Contractors cannot be expected to be experts in design; rather, they are experts in the means and methods of construction.

In addition to reviewing the project plans and specifications carefully, the contractor will also want to review carefully any supplemental documentation provided by the owner. This may include old plans and specifications, boring information, environmental audits, and any inspection reports or other documentation provided by the owner. The objective of these reviews is similar to the reviews performed by the owner and the designer. The contractor is trying to discern the type, location, and extent of contamination that might be discovered during construction.

Inspecting the site before providing a price for performing the work is also an essential part of the contractor's research effort. The purpose of this visit is to become familiar with the visible characteristics of the site. If contamination is present and visible, then one of the objectives of the site visit will be to note any visible impediments to completing the scope of contract work that applies to this contamination. For example, the contractor will want to evaluate access and the availability of utilities. Usually, the contract will make the contractor responsible for knowing and accounting for anything that could have been determined from a site visit.

After having become thoroughly familiar with the plans, the specifications, the site, and other available information, the next planning exercise to be performed by the contractor is to figure out how the contract work will be accomplished with regard to contamination. Once the contractor understands how the work will be accomplished, a cost estimate can be developed. This cost estimate will not only account for the anticipated costs to perform the contract work, it will also address the risks represented by the contract language.

Execution

As discussed earlier in this chapter, project management consists of two activities. These activities are planning and execution. A project's planning phase is essentially complete when the work begins, though there may be some overlap. Just as the major planning efforts can be divided among the major players, the major execution actions can be divided as well.

The owner's role

The owner can take various steps to help assure the successful completion of a project where contamination is present or possible. Each of these will be addressed roughly in chronological order.

Design and constructability reviews. The project design process will usually profit from a series of well-timed checks or reviews. These types of reviews take two forms. The first type of review is often called a *design review*. It is performed early in the design process. The objective of a design review is to evaluate the basic approach to the design of the project. This might include an evaluation of the basic feasibility of the design, an evaluation of whether it meets the basic design objectives, and consideration of alternative approaches to the design. A design review should be done as early as possible, before a lot of time has been expended preparing contract drawings and before changes to the design have a serious impact on the schedule of the design process.

A *constructability review* is performed later in the design process, after the designer has begun to prepare contract drawings. The objective of a constructability review is to identify errors, omissions, ambiguities, other problems with the plans and specifications, and more economical or alternative approaches to systems before the bid package goes to the contractor for bidding.

Both designers and owners often assert that they perform design and constructability reviews in house. In reality, the owner's team is often saddled with far too many responsibilities and may have neither the time nor the technical expertise to conduct a detailed design or constructability review. The designer's staff is often much too close to the design to be able to perform an effective and objective review. Consequently, it is often advisable to secure the services of a third party to conduct design and constructability reviews. For projects where contamination is an issue, the third party selected to do design and constructability reviews should have appropriate knowledge and experience with the kinds of contamination that are expected to be encountered. Also, the requirement to submit its design to a third party should not be sprung on the designer at the last minute. This requirement should be specifically addressed in the designer's agreement. It should also be scheduled into the design phase of the project. The schedule shown in Fig. 3.10 provides for design and constructability reviews. By scheduling in advance, adequate time can be set aside for thorough reviews.

Bonds. Three basic types of bonds may be required on a construction project.

1. *Bid bond.* A bid bond is typically provided with a contractor's lump-sum bid when it is submitted to the owner. It essentially provides a monetary guarantee to the owner that a contractor will accept the award of a contract. Should the lower bidder refuse to perform prior to award, then the owner may seek recovery of the cost differential between the low bid and the second lowest bid from the bonding company that has underwritten the bond.

2. *Payment bond.* A payment bond provides for the payment of subcontractors, materialmen, and suppliers in the event that the contractor fails to make payments. This type of bond allows the subcontractor to seek payment from the surety, rather than to file a lien on the project.

3. *Performance bond.* A performance bond is a bond that theoretically provides for the completion of the project should the contractor default on the contract. Payment and performance bonds are often issued together.

Bonds are often thought of as insurance policies. They insure that the contractor will perform the contract at the price bid and that subcontractors, materialmen, and suppliers will be paid for the work they perform. The protections offered by a bond, particularly a performance bond, may be of particular importance in situations where the contractor will be handling the abatement or remediation of hazardous materials. A bond will help alleviate some of the discomfort associated with the nightmare scenario of a contractor defaulting in the midst of cleaning up and disposing of hazardous materials. In such a situation, the owner could be left with an open, unprotected, and contaminated site along with the contractor's potentially contaminated equipment. There may also be serious questions concerning whether the contractor properly disposed of the hazardous waste removed from the site. In this scenario, the surety holding the bond would be expected to step in and complete the work.

Owners, however, should be fully aware of the protections actually offered by a bond. Figure 3.11 provides the actual text of a performance bond. Note that the surety is only obligated to the dollar amount of the bond. The dollar amount of the bond will often equal the contract value. This amount is not reduced as the contract work is performed and paid for. It does, however, represent the maximum contribution the surety will make toward completing the project. Consequently, should the contractor leave the project in such a condition at the time of default that the correction of deficiencies and completion of the contract work exceeds the dollar amount of the bond, the owner may be responsible for the excess costs. Also, owners should recognize that both contractors and their sureties may dispute the propriety of the owner's default termination of the contract. In such cases, the owner may have to arrange and pay for the completion of the work and seek recovery of the resulting costs from the contractor and its surety.

Another benefit of requiring the contractor to be bonded is that the owner gains the benefit of the surety company's evaluation of the contractor's financial health and prospects. Though even sureties are sometimes surprised, they generally have greater access to a contractor's financial records and, thus, are often better able to evaluate the contractor's financial health. Sureties may also have required the ownership of the contractor to personally guarantee the value of the bond. This may serve as an added incentive for a troubled contractor to finish a bonded project.

The additional security offered by a bond comes at a price. Bond premiums are generally set at 1 to 2 percent of the bond value.

Pre-bid meetings. It is in the owner's and the contractor's best interest to assure that the contractor is as knowledgeable as possible about the project, existing conditions, and other information germane to the contractor's performance of its work. This is particularly true when contamination is or may be present on the site. One of the things the owner can do to help the contractor understand all the requirements of the project, and in particular, problems that may be encountered, is to schedule a pre-bid meeting. The primary objective of a pre-bid meeting is to provide contractors interested in performing the work with an opportunity to ask questions, point out problems with the contract documents, and hear the owner's description of the project and some of the problems that will be faced. The pre-bid meeting is also an excellent forum in which to discuss potential or existing site contamination. It also gives the owner or designer an opportunity to discuss unusual or innovative features of the design. Demanding or stringent requirements can be pointed

Performance Bond

KNOW ALL MEN BY THESE PRESENTS that we, _____, _____ as Principal (hereinafter called the "Principal") and _____ as Surety (hereinafter called the "Surety"), are held and firmly bound unto _____ (hereinafter called "the Obligee"), in the sum of _____ DOLLARS ($_____). for the faithful performance of the Contract defined below, lawful money of the United States of America, for the payment of which the Principal and the Surety bind themselves and their respective successors and assigns, jointly and severally, firmly by these presents.

WHEREAS, the Obligee is a "contracting body" under provisions of the "Public Works Contractors' Bond Law of 1967," as amended (the "Act"); and

WHEREAS, the Principal intends to enter into an agreement dated _____, 199____ (the "Contract") with Obligee for _____, a copy of which Contract is by reference made a part of the Bond; and

WHEREAS, the ACT requires that the Principal shall furnish this Bond to the Obligee;

NOW, THEREFORE, the terms and conditions of this Bond are and shall be that if the Principal well, truly and faithfully shall comply with and shall perform the Contract in accordance with its terms, at the time and in the manner provided in the Contract and if the Principal shall satisfy all claims and demands incurred in or related to the performance of the Contract by the Principal, and if the Principal shall indemnify completely and shall save harmless the Obligee or any of its members, directors, officers, employees and agents from any and all costs and damages which the Obligee or any of its members; director, officers, employees and agents may sustain or suffer by reason of the failure of the Principal to do so, and if the Principal shall reimburse completely and shall pay to the Obligee any and all cost and expenses which the Obligee or any of its members, directors, officers, employees and agents may incur by reason of any such default or failure of the Principal if the Principal shall remedy, without cost to the Obligee, then this Bond shall be void, otherwise, this Bond shall be and shall remain in force and effect.

The foregoing, however, is subject to the following further provisions:

1. This bond shall be construed in accordance with the laws of the Commonwealth of Pennsylvania. The Principal and the Surety agree that exclusive jurisdiction and venue for any litigation concerning this Bond and the transactions contemplated shall exist in the Chester County, Pennsylvania, Court of Common Pleas. The Principal and the Surety consent to such jurisdiction and venue and agree that all service of process including any instrument to institute suit, shall be effective if served in accordance with Pennsylvania law.

2. The surety hereby waives notice of and consents (a) to all alterations or amendments to the Contract and (b) to all extensions of time for performance of the Contractor or other forbearance; and the Surety agrees that its obligations under this Bond shall not thereby be released or affected in any manner.

3. The Surety shall not be liable under this performance Bond to the Obligee in the aggregate in excess of the sum above stated.

SIGNED and SEALED this _____ day of _____, 199_____.

_____
(Principal)

ATTEST:

_____ BY:_____

BY: _____
(Surety)

_____
Attorney-in-fact (Attach power of attorney)

Figure 3.11 Performance bond.

out, particularly if these requirements differ from what the contracting community in the region might be used to. Though the owner has no obligation to make such a presentation, it may serve as a means to prevent surprises and reduce sources of conflict.

The date, time, and location of the pre-bid meeting should be set forth in the bid documents. A pre-bid meeting should be mandatory. In other words, contractors may not bid the project if they fail to attend the pre-bid meeting. Minutes of the pre-bid meeting should be kept and provided to all bidders as an addendum to the bid package.

Partnering. Over the years, the construction industry has become increasingly litigious. With this increasing litigiousness there has also arisen the feeling that the many participants to a construction project are finding it increasingly difficult to form effective project teams capable of solving even the simplest project problems without dissolving into bitter disputes. In an effort to address this problem, techniques have been developed both to try to decrease the amount of litigation and to foster the development of stronger project teams. One of these team-building techniques is known as *partnering*.

Though it can be said that participants on construction projects have partnered for years, the term *partnering* as used here means something more specific. Partnering is the formal process by which the parties involved in a construction project are brought together in a single room at one time with the express purpose of forming a project team. These get-togethers are known as *partnering sessions*. The first session is generally held shortly after the contract is executed, but before the contractor's work begins. Follow-up sessions are generally held as needed throughout the remainder of the project.

The first step toward providing for a partnered project is simply to decide to partner. Owners may actually provide for partnering in the contract. The following is the partnering clause that appeared in the contract for a large bridge project.

> *104.01 Intent of Contract Documents.* The intent of the contract documents is to provide for the construction and completion of the work described to the satisfaction of the Department. The Contractor shall furnish all labor, materials, equipment, tools, transportation services, and supplies required to complete the work in accordance with the contract documents.
>
> a. *Covenant of Good Faith and Fair Dealing.* This contract imposes an obligation of good faith and fair dealing in its performance and enforcement.
>
> The Contractor and the Department, with a positive commitment to honesty and integrity, agree to the following mutual duties:
>
> 1. Each will function within the law and statutes applicable to his duties and responsibilities.
> 2. Each will assist in the other's performance.
> 3. Each will avoid hindering the other's performance.
> 4. Each will proceed to fulfill its obligations diligently.
> 5. Each will cooperate in the common endeavor of the contract.

b. *Voluntary Partnering.* The Department intends to encourage the foundation of a cohesive partnership with the Contractor and its principal subcontractors and suppliers. This partnership will be structured to draw on the strengths of each organization to identify and achieve mutual goals. The objectives are effective and efficient contract performance and completion within the budget, on schedule, and in accordance with the Contract Documents.

This partnership will be bilateral in makeup, and participation will be totally voluntary. Cost associated with effectuating this partnering will be agreed to by both parties and will be shared equally.

To implement this partner initiative prior to commencing with the work in accordance with the requirements of Subsection 108.02 and prior to the preconstruction conference, the contractor's management personnel and the Engineer will initiate a partnering development seminar/team building workshop. Project personnel will make arrangements to determine attendees at the workshop, agenda of the workshop, duration, and location. Persons required to be in attendance will be the Department key project personnel; the contractor's on-site project manager and key project supervision personnel of both the prime and principal subcontractors and suppliers. The project design engineers, FHWA, and key local government personnel will also be invited to attend as necessary. The contractors and the Department will also be required to have Regional/District and Corporate/State level managers in attendance.

Follow-up workshops may be held periodically throughout the duration of the contract as agreed by the Contractor and the Department.

The establishment of a partnership charter on the project will not change the legal relationship of the parties to the contract, nor relieve either party from the terms of the contract.

As indicated in this specification, the first partnering session for the project is usually held before the preconstruction meeting and before the project construction work has begun. Logistically, partnering is usually performed at a location remote from the site. It is not performed, however, at the offices of the owner, the contractor, or some other contracting party. Instead, it is performed at a neutral location. A hotel or some other conference facility is often chosen. Those in attendance consist of all the significant project participants. This includes the owner and its staff and the contractor and its staff. As indicated by the quoted partnering contract clause, attendees may also include the designer and its staff, major subcontractors and suppliers, and even local government officials and regulatory bodies. Adjacent landowners may also send representatives.

Though the parties may conduct the partnering session themselves, a partnering facilitator is often used to help the participants get the most out of the session. In addition, it is sometimes awkward to have one of the parties to the contract act as the facilitator in the session. For this reason an independent third party is often preferred. When selecting a facilitator for a project where contamination is expected to be found or is known to exist, investigate whether the facilitator has any kind of experience with these kinds of projects. Also, since many facilitators hail from academia or are purely management consultants, find out whether the facilitator has any experience at all with construction. A facilitator with experience in construction may be helpful because then he or she will be familiar with the relationships on a construction project and the events that can harm those relationships. It may also be helpful because the facilitator will be familiar with common construction terminology and can speak the language of the participants.

The facilitator will provide a number of services other than simply attending the session. Though administrative functions such as arranging for the meeting rooms, scheduling the session, and inviting the participants are usually handled by others, the facilitator is responsible for actually planning the facilitation session. Usually, the facilitator prepares a small handout for distribution at the session. This handout serves as a workbook, notebook, a schedule, and an agenda for the session. Prior to the session, the facilitator should learn something about the parties who will be attending and the project itself. If they are not already known, the facilitator should be briefed on the problems likely to be encountered on the project. These are the problems that are most likely to affect the cohesiveness of the project team.

Payment for the partnering session is usually split by the owner and the contractor.

The primary objective of the partnering session is team building. To this end, the facilitator leads the participants through a series of exercises designed to break down the barriers to forming an effective team and then helps the participants to form such a team. The culmination of the partnering session is the development and signing of the partnering charter. The charter is signed by all in attendance at the partnering session. The following is the text of a charter from a hazardous waste remediation project.

> We the partners of the _____ project agree to improve the working relationships and project environment by joining together to proceed from this point on, as a cohesive team. Our common goal is to complete the technically complex remediation of the site in a safe, environmentally acceptable manner. Members of the Partnering team will have the courage to resolve past problems in a fair and equitable manner and then focus their energies on the future success of the project. They will deal with each other in a reasonable, open, trusting and professional manner. In that spirit, we are committed to the following objectives:

Communication	Performance
1. Communicate problems, at all levels, openly and as early as possible.	1. Restore the site to an environmentally safe and attractive condition.
2. Resolve problems and make decisions at the lowest possible level in a timely manner.	2. Execute and administer the contract so that all parties are treated fairly and equitably.
3. Maintain a professional atmosphere of mutual respect and resolve conflicts immediately.	3. Avoid preventable delays on the project.
4. Talk before we write.	4. Eliminate formal disputes.

Communication	Performance
5. Develop a periodic feedback process on the partnership's communication program.	5. Perform a high-quality activity to enable all participants to take pride in undertaking an extremely difficult task.
6. Respond to community concerns in a timely and professional manner.	6. Promote a sense of ownership on the part of every participant.
	7. Complete the work in a timely and effective manner in accordance with contractual and environmental requirements.

The following charter is more generic.

(PROJECT NAME)_____
PARTNERING AGREEMENT

We are the partners in the (Project Name)_____. We commit to an open, trusting, and cooperative relationship and will pursue the following goals in good faith:

 I. Timely solutions to problems and resolutions of disputes at the lowest possible level.
 II. Timely project completion.
 III. Open communications; agree to meet weekly to discuss schedules and major issues. Agree to prepare agendas for each meeting and minimize the durations of the meetings.
 IV. Overall outstanding performance report.
 V. Commitment to subcontractor coordination.
 VI. Safe workplace for workers and the public.
 VII. Reasonable contractor profit.
 VIII. Quality product.
 IX. No unresolved disputes, claims, or litigations at end of project.
 X. Timely and accurate submittals by contractor and subcontractor and review by the State.
 XI. Clearly defined organizational structure and delegation of authority by both parties.
 XII. Pay close attention to interpersonal relationships and strive to create a pleasant work atmosphere.
 XIII. Commit to a monthly review of the partnering process by all personnel. The project manager and project engineer will conduct the review using the evaluation form, will summarize and discuss the results, and will take appropriate actions to address any areas of concern.

Signed:

_____ _____

_____ _____

Both charters start with what is known as a mission statement. A *mission statement* is a basic statement of the project team's primary objective or mission. The mission statement is followed by a list of goals, objectives, and philosophies that describe how the parties intend to accomplish the overall mission objective. All participants to the partnering session sign the charter.

Partnering is not accomplished in one session. Though the typical partnering session lasts two days, this is not enough time to form a lasting project team, particularly on projects that face significant problems. To assure the success of the team, the project participants must actually act as they agreed in the partnering session. In addition, it is often necessary to have follow-on partnering sessions to work on areas where the partnership is failing, to deal with specific issues, and to reinforce the goals and objectives established in the charter.

The goal of partnering is simple: It is to establish the working relationships which will exist on the job. While the contract establishes the legal relationships, the partnering charter defines the day-to-day working relationships.

Response time. One of the chief responsibilities of the owner and its staff, perhaps assisted by the project designer, is to review and respond to questions, submittals, and problems identified by the contractor. It is essential to the success of the project and the owner's relationship with the contractor that these questions, submittals, and problems be answered or addressed as quickly and as completely as possible. There are a number of things an owner can do to facilitate the process of responding to the contractor's questions. The first is to establish good oral communications and provide forums in which problems can be identified and discussed. On large projects it may be a good idea for the owner's representative and the contractor's superintendent or project manager to meet informally several times a week to discuss items of interest. In addition, a weekly or at least monthly project meeting where minutes are kept is a necessity. The focus of this meeting should be on problem resolution and planning upcoming events. An "Action Items Control Log" should be maintained to track the resolution of ongoing problems. This control log is a particularly useful and successful approach to documenting and tracking problems in the project. An Action Items Control Log procedure is included in App. D.

The procedure and forms describing how to use an Action Items Control Log are often provided in the project manual. The *project manual* sets forth the project's basic administrative procedure. These administrative details are not usually provided in the contract documents. Figure 3.12 is a listing of the typical components of a project manual. The exact components of the project manual will vary based on the administrative systems and procedures to be used on the project. The manual should also include samples of the administrative forms to be used on the project. Both blank and properly completed forms should be provided.

Most significant items requiring the owner's action should be handled in writing, with a system developed for organizing, reporting, and tracking the resolution of these items. For questions, the most common mechanism for

> 1. Cover page
> 2. Table of contents
> 3. Description of the project, participants, and purpose of the project manual
> 4. Partnering charters signed for the project
> 5. A list of the names, titles, and authority of persons on the project, including home phone numbers, as appropriate. This will be useful on projects where hazardous materials are present or expected to be discovered.
> 6. Change order procedures
> 7. Scheduling procedures
> 8. Submittal procedures
> 9. Project progress meetings organization and procedures
> 10. Safety procedures
> 11. Project daily log and diary procedures
> 12. Inspector's daily report procedures
> 13. Correspondence requirements, formats, and addresses of project participants
> 14. Request for Information (RFI) procedures
> 15. Action items control log procedures
> 16. Telephone log procedures
> 17. Project filing procedures
> 18. Sample forms—blank and properly completed
> 19. Procedures for discovery of hazardous materials

Figure 3.12 Components of a project manual.

identifying and tracking is called a *Request for Information* (RFI). RFIs are prepared by the contractor or subcontractor on a standard form. The form includes a unique number for the RFI, an indication of the party posing the question, the date the question is posed, and space for the question itself. Though the owner may also use the RFI system to pose questions to the contractor, it is best to keep the owner's RFIs separate from the contractor's RFIs. This includes tracking them separately. The contractor submits the original of the RFI to the owner. A copy is retained by the contractor and filed sequentially with the other RFIs. It is also listed in the RFI log. Each RFI is listed in the RFI log by number, date submitted, and description. A column should also be reserved in the RFI log indicating when the owner's response is received. A typical RFI form is shown in Fig. 3.13.

When the owner receives an RFI, it should be logged in. The owner's RFI log should also list each RFI numerically, by date, and description. The owner should strive to answer the RFI as quickly as possible. If the questions can be answered the same day, the owner should do so. Often the RFI form will include the date by which the RFI must be answered in order not to affect the project adversely. This is a helpful addition, since it allows the owner to prioritize efforts. If the RFI requires input from the designer or others, it should be forwarded to these other parties as quickly as possible.

REQUEST FOR INFORMATION

RFI No.: _____ Date: _____

Project No. _____ Time: _____

INFORMATION REQUESTED: Reference: Plan Sheet(s)_____
                             Specifications(s):_____

Initiator: _____ Phone: (___)_____
Date response is needed: _____ Fax: (___)_____

Complete form above and transmit to Resident Engineer for action. Keep a copy for your records.

RFI received: _____(DATE)_____(TIME)_____(BY)
Action Item Log No.: _____ Action Party: _____
RFI Transmitted to Action Party: _____
RFI Received by Action Party: ___(DATE)___(TIME)___(BY)
                              ___(DATE)___(TIME)___(BY)

RESPONSE: Response Date: _____

Response Transmitted to Initiator: _____
                                   (DATE)   (TIME)   (BY)

Figure 3.13 Request for Information (RFI) form.

The RFI form usually contains space for the owner's reply. This space should include space for the reply itself, the date of the reply, and the name of the party providing the reply. When it is returned by the owner to the contractor the date of return should be indicated. The contractor should also log in the date of the owner's reply. This system as described is for a paper-based RFI system. If the parties are linked electronically, the system will work similarly, but there will be no need to exchange paper.

The system used to track submittals would be similar to the system used to manage RFIs. A standard submittal form or cover sheet should be used, and submittals should be numbered, dated, and described as they are logged out and in by each party in the chain. A typical submittal form is shown in Fig. 3.14. The log forms for submittals should be flexible to allow for multiple submittals, as it is sometimes necessary for the contractor to make multiple submittals to receive approval. The owner may require more time to review and return submittals, particularly if they are voluminous. Depending on the type and magnitude of the submittal, turnaround time may vary from a few days to a few weeks. It is often helpful to indicate the time required to review and approve submittals in the contract. A typical allowance for reviewing submittals might be two weeks (10 working days). Additional time might be built in for larger submittals. Some contracts allow the owner an additional day of review time for each page of material submitted. Few submittals should require more than one or two months to turn around. If reviewing and returning the submittals is going to take longer than expected, the contractor should be notified.

There are a number of ways to track the identification of problems on a project. The complexity of the system will depend on the size of the project and the number of problems. Any system used should reflect the requirements of the contract and realistically reflect the way problems should be addressed on the project. For example, consider a backhoe operator who discovers potential contamination. A system for documenting and tracking this discovery might work as follows. Initial notification of the problem is oral. The equipment operator notifies the superintendent. The superintendent in turn notifies the owner or the owner's representative. Both the superintendent and the owner should probably note the discovery in their daily logs or diaries. These diary entries should focus on such basic information as the time and place of discovery, the party discovering the potential contamination, a description of the conditions discovered, and any information known regarding injuries or other health and safety-related information. The same day the oral notice is provided, either the contractor or the owner may provide handwritten documentation of the oral discussions via a "speed memo" or some other form of written communication. Such memoranda should also confirm any oral directive provided concerning how the project team is to address the discovery. This includes directives to stop work and actions taken to contain the potential contamination. Ultimately, most contracts require the contractor to provide the owner with written notification within a specific period of time. This period of time is often short, probably less than one week.

LETTER OF TRANSMITTAL

TO: Company Name and Address

DATE: _____ JOB NO.: _____
ATTENTION: _____
RE: _____

WE ARE SENDING YOU:
- ☐ Under separate cover via _____ the following items:
- ☐ Attached
- ☐ Shop Drawings
- ☐ Prints
- ☐ Plans
- ☐ Samples
- ☐ Specifications
- ☐ Copy of Letter
- ☐ Change Order

COPIES	DATE	NO.	DESCRIPTION

THESE ARE TRANSMITTED as checked below:
- ☐ For approval
- ☐ For your use
- ☐ As requested
- ☐ For review and comment
- ☐ FOR BIDS DUE _____, 19___
- ☐ Approved as submitted
- ☐ Approved as noted
- ☐ Returned for corrections
- ☐ _____
- ☐ Resubmit _____ copies for approval
- ☐ Submit _____ copies for distribution
- ☐ Return _____ corrected prints
- ☐ PRINTS RETURNED AFTER LOAN TO US

REMARKS: _____

COPY TO: _____ SIGNED: _____

Figure 3.14 Standard submittal form.

Once notified, the owner has to provide a response. This response should be made as quickly as possible, particularly when contamination is involved. In addition, the contract may stipulate the amount of time available. Though the owner may respond to the contractor's notice orally, a written response should ultimately be provided. If the discovery will result in the need for a modification to the contract, the owner should inform the contractor that a change is to be made to the contract and that a proposal is required. At this point it is advisable to establish a separate filing and tracking system for the issue. The owner's request may be given a *Change Order Request* (COR) number. The contractor's proposal may be given a *Change Order Proposal* (COP) number. Once the contractor's proposal has been negotiated and agreed to, a change order can be executed. Change orders are also usually numbered consecutively. CORs, COPs, and change orders are all logged and filed separately.

In summary, it is essential for the owner and the owner's representatives to respond quickly to questions posed, submittals made, and problems identified by the contractor. Though numerous administrative systems, both paper-based and electronic, facilitate this process, there can be no substitute for prompt attention by qualified personnel.

Monitoring. Another of the owner's major responsibilities on a project is to monitor the contractor's performance and the completion of the work. On a project where contamination is likely to be found or is known to exist, the owner will probably be concerned with more than simply assuring that the contractor's work is completed in accordance with the contract documents. The owner will also want to be vigilant and assure that the contractor's forces recognize and respond properly to the presence of contamination. In addition, the owner will want to ensure that the contractor's practices do not themselves pollute the site and the surrounding environment. Fuel spills should be identified and cleaned up properly. Construction waste and debris should not be buried and burned on the site, at least not without proper permits from appropriate government authorities. Appropriate sediment controls should be established to prevent the fouling of local bodies of water. These sediment control methods will be established by contract specification or by government regulation.

In order to accomplish its monitoring responsibilities, the owner will employ inspectors and others who are well trained and experienced with the work to be inspected. In the case of contamination, inspectors and others in monitoring roles should be trained or have experience in the recognition of likely contaminants. They should also know what actions to take when contaminants are found.

One problem too often faced by owners is a contractor who refuses to proceed as directed. For example, on a bridge contract, the painting subcontractor refused to comply with the regulations governing the removal of lead-based paints and primers from bridge girders. The owner's inspection staff recognized that the painter was proceeding with the work inappropriately, and that there was a potential that workers and others nearby were breathing in unsafe levels of lead. There was also the potential that the soils underneath the bridge were becoming contaminated with lead. The inspector notified the painter orally of the problem. The painter ignored the inspector and continued with the unacceptable paint-removal activities.

The approach to encouraging a recalcitrant contractor to conform with the contract requirements is one of escalation until the appropriate amount of leverage can be applied to force the contractor to comply with the contract requirements. The speed with which this escalation takes place is a function of the seriousness of the problem, whether or not the problem is recurring, and the owner's confidence in its position. In the case of the painter, because the painter is endangering not only its workers but others, and is also potentially spreading the contamination, escalation should be swift and without hesitation, and direction should be given clearly and forcefully. If the inspection staff is ignored by the painting subcontractor, it should escalate the problem immediately to a supervisor (often the owner's resident engineer or project manager), who will immediately contact his or her counterpart in the general contractor's organization. In this case, the direction given will be for the general contractor to stop the paint-removal work until the painter can bring operations into conformance with the contract and applicable regulations. Usually, in a case such as the one described, this step will be sufficient to stop the paint-removal work and force compliance. The owner's representatives should send written confirmation of the stop work order, and the events that transpired up to the issuance of the stop work order should be recorded in the daily logs and diaries of those involved.

Resolution may be more difficult in situations where the threat to health and safety is not so immediate or obvious and the contractor's actions are not in direct contradiction of the contract requirements, but more an issue of contract interpretation. In cases such as these, the owner's initial actions may not be sufficient to force compliance immediately. A common example of this type of problem relates to the contractor's level of effort on a project. The owner or the owner's representatives may conclude that the contractor has not mobilized sufficient personnel and equipment to proceed with performance of the contract work at an adequate pace. The owner may have come to this conclusion based on experience or other indicators such as delays to the project schedule.

The following are the levels of escalation through which resolution of this problem may proceed.

The first step is oral discussions between the owner's representative and the contractor's project superintendent. These discussions should be kept on a purely professional basis. The owner's representative should convey the owner's concerns, and the evidence that supports these concerns. The owner's representative should also convey the owner's desire that the contractor take action to halt the continued delays to the project. The owner's representative should document this conversation in his or her daily diary.

If the contractor fails to respond, the next step is to document the owner's concerns in writing to the contractor. The problem may be discussed in the periodic project meetings. The minutes of these meetings document the problem. A letter should also be sent from the owner's representative to the contractor's superintendent or project manager. The letter should recount the previous oral discussions and the fact that the problem has not been corrected since these discussions took place.

What happens next will be a function of the contractor's reaction. One response would be silence but with an increase in personnel, which would resolve the problem. Another approach might be a written response explaining the situation and describing the corrective action to be taken if any is required. This response also indicates resolution of the problem, though the owner will have to monitor the project to ensure that the contractor's promises are kept.

A third possible response from the contractor might be to attempt to blame the problems of the project on others, the weather, or the owner. The contractor might even request a time extension or added compensation. It is this response that will cause further escalation. First, the owner must review the contractor's letter. If the contractor's allegations are correct, then the owner must take action to alleviate the problems hampering the contractor's progress. If the contractor's position is unfounded, then the owner must respond in writing and in detail, describing why the contractor's position is incorrect. The letter should conclude with continued direction to the contractor to correct the problem.

At this point, and perhaps long before this point is reached, meetings between the owner and contractor should be conducted to discuss the problem, its causes, and possible resolutions.

If the owner continues to feel strongly that it is correct and the contractor remains recalcitrant, then escalation to the next level is appropriate. This next level might consist of a number of different actions or a combination of many actions, including the following:

1. Holding a special partnering session focusing on the issues that cannot be resolved.

2. Escalating the problem to the next management level. For the contractor that next level might be a vice president or the owner of the company. For the owner that might be someone higher in the corporation or bureaucracy. On federal government jobs, this next level might be the Contracting Officer.

3. If there is a specific dispute that is the source of the problem and the contract documents have provided for the use of a disputes review board (DRB), then the dispute might be brought to the DRB in an attempt to resolve it.

Having escalated the problem to the highest level and made every attempt to resolve disputes, if the owner is still unsatisfied with the contractor's response, then it is time to look to the contract for available enforcement tools. These tools might include withholding payments or bringing other contractors onto the project to perform the work the contractor refuses to perform. If the project schedule shows that the project is delayed and the work will not be complete by the contract milestone or completion dates, and the owner determines that the contractor is responsible for these delays, then the owner may begin assessing liquidated damages. If the problem is a particular person or subcontractor, the owner may take action to have the person or subcontractor removed from the project site.

If the enforcement tools or sanctions provided by the contract are insufficient to convince the contractor to take action on the owner's complaints, the owner must decide whether to escalate to the next and most drastic step. This step is termination. If the owner concludes that the consequences of removing the contractor from the site are outweighed by allowing the contractor to continue work, then termination becomes a viable strategy.

Termination of a contract is a drastic action. It is generally undertaken only as a last resort, when all other options have been exhausted. Consequently, it should not be threatened unless the owner is willing to follow through on the threat. The process by which a contract is terminated for default is spelled out in detail in the contract. The owner must follow these procedures precisely. For example, the contract generally provides for some period after the contractor is notified in writing that it is in default before the contract can be terminated. This period is often three to seven days in length. It is provided to give the contractor an opportunity to "cure its default." If the owner terminates the contract before the cure period is exhausted, the default termination could ultimately be determined to have been wrongful.

One question often asked on projects where the contract requires the contractor to be bonded is when to notify the bonding company of problems with the contractor's performance. The answer to this question depends on the circumstances. For example, if the cause for concern relates to the intentional flouting of regulations concerning the handling, cleanup, or disposal of hazardous materials, and the contractor is endangering others or spreading the contamination, then notification of the bonding company may need to be immediate. Notification should always be in writing, including a detailed explanation of the reasons for concern. Thereafter, the bonding company should be sent copies of all correspondence. In addition, the contractor should be sent copies of all correspondence to the bonding company. In less extreme situations, the bonding company should be made aware of any problems when situations where contract remedies and sanctions are being considered, particularly if payment will be withheld. The withholding of payments can affect the contractor's financial stability, an issue of prime concern to the bonding company.

Related to the issue of termination, owners should be careful not to overpay the contractor for work performed. The withholding of retainage is one way to help control this problem. If the contractor is performing defective work, however, retainage can be eaten up quickly. It has been argued that owners have a responsibility to the surety not to overpay contractors for work performed. When the contractor is overpaid, there may be less money available in the contract than is needed for the surety to complete the contract work.

The contractor's role

The contractor is the primary executor of the contract's requirements. There are many tools available to the contractor to enhance its ability to execute the contract requirements. With regard to the discovery of contamination, there are at least three areas where good management practices can assist the contractor in accomplishing the project requirements. The first of these areas is training, which is discussed in detail in Chap. 5.

Scheduling. A second area where good management practices can assist in dealing with contamination on site is scheduling. Many contracts require the contractor to prepare a project schedule. Though such provisions may be quite detailed (see, for example, the provisions given in Chap. 4), it is usually up to the contractor to fulfill not only the requirements of these specifications, but the intent as well. A project schedule that is complete and thoughtfully developed will not only assist the contractor in managing the basic contract work, but will also help in dealing with the inevitable problems encountered on construction projects, not the least of which is the discovery of contamination.

Construction schedules come in two basic forms. The first, oldest, and perhaps most common form is called a *Gantt* or *bar chart*. The name Gantt comes from Robert Gantt, who developed the bar chart approach to scheduling in the early 1900s. The more common term today is bar chart. Figure 3.15 shows an example of a bar chart schedule. The left column consists of a list of activities. Across the top of the schedule is a calendar. Each bar in the schedule corresponds to a particular activity. The beginning of each bar corresponds to the planned start date of the activity. The end of the bar corresponds to the planned finish date.

A bar chart schedule has several strengths. The first and primary advantage is its simplicity. Almost anyone, even those with no experience in construction or scheduling, can understand a bar chart schedule. This makes a bar chart schedule a powerful method for conveying construction scheduling information to upper management, administrative boards, the public at large, and other organizations and individuals who are not normally involved with construction. Note also that a bar chart schedule is easy to develop: No special training or expertise is required.

Unfortunately, a bar chart schedule also has several weaknesses. The primary disadvantage is that a bar chart schedule does not show relationships among the activities in the schedule; the schedule does not indicate whether the start of one activity depends on the completion of another. The second disadvantage relates to the fact that a bar chart schedule is a graphic schedule. It is drawn. Consequently, the level of detail is limited by the size of the paper. A total of 20 or 30 activities is the most one could hope to get on a single sheet of $8\frac{1}{2} \times 11$ paper. Though using a larger sheet of paper or more sheets allows the inclusion of more activities, this begins to erode the strengths of bar chart schedules, making them more difficult to prepare and reproduce and harder to read and understand.

Because relationships are not shown and because as a practical necessity the work must be shown in a summary fashion, it is often difficult to figure out how the addition of work will affect the schedule and to update the schedule to show

Activity ID	Activity Description	Orig Dur
300	STORM SEWER III	30
40	NORTH LOT PHASE #1	15
305	ROUGH GRADE III	15
390	SLAB III	15
110	FINE GRADE I	8
210	FINE GRADE II	9
120	SUB BASE I	9
160	LANDSCAPE I	10
310	FINE GRADE III	9
395	CURB III	9
170	SIDEWALKS I	20
190	CURBS & SLABS I	10
280	FENCE II	10
130	BLACK TOP BINDER I	9
220	SUB BASE II	9
260	LANDSCAPE II	10
270	SIDEWALKS II	20
60	CURBS	20
20	TRAFFIC CONTROL SIGNALS	20
30	GUARD & BUS SHELTERS	20
180	FENCE I	10
50	NORTH LOT PHASE #2	15
140	BLACK TOP TOP I	9
230	BLACK TOP BINDER II	9
320	SUB BASE III	9
360	LANDSCAPE III	10
380	FENCE III	10
240	BLACK TOP TOP II	9
330	BLACK TOP BINDER III	9
150	PAVEMENT STRIPING I	3
370	SIDEWALKS III	10
340	BLACK TOP TOP III	8
250	PAVEMENT STRIPING II	3
350	PAVEMENT STRIPING III	3
400	PROJECT COMPLETE	0

Figure 3.15 Bar chart schedule for an airport employee parking lot.

the status of the project. Thus, in situations where contamination is discovered, it is difficult to determine how the discovery will affect the project schedule.

In an effort to correct some of the deficiencies of bar chart schedules and also to facilitate computerization, a new kind of scheduling technique was developed. This technique, known as the *critical path method* (CPM), was developed in the late 1950s by Dupont and the Rand Corporation. Over the years this scheduling system has developed and spread to the point that it is now common in the construction industry.

At its heart, a CPM schedule is a logic network. In a logic network, every activity in a project is listed and the relationship of each activity to other schedule activities is identified. Figure 3.16 illustrates a network diagram for a small project. This network diagram was drawn in the *activity-on-node* (AON) format. The work associated with each activity is described at a node. These nodes are then connected to other related activities. In Fig. 3.16, the logic connections are represented by arrows. Another network diagram format is known as *activity-on-arrow* (AOA). In this format the work activities are described on the arrows. The nodes simply show how each activity is related. Though for years the AOA format was by far the most common and is still specified by some owners, it is being eclipsed by the AON format. This is primarily because AON networks use fewer activities to describe the same work. Perhaps as many as 20 to 30 percent fewer activities are required. These savings occur for two reasons. The first is that AON schedules do not require the use of special, zero-duration activities known as *dummies* or *restraints* to supply all the logic required by the schedule. The second reason is that the AON networking technique allows a greater variety of relationships. In an AOA network, only finish-to-start relationships are allowed. In other words, the start of all but the first activities must be preceded logically by the finish of the activities that logically preceded them. Other types of relationships are possible on AON schedules, including finish-to-finish relationships and start-to-start relationships. For these reasons, many construction professionals prefer AON schedules. Because contractors are often charged with responsibility for preparing the project schedule, AON schedules have become more prevalent. This development has not gone unnoticed by scheduling software developers. While most scheduling software will support the preparation of AOA schedules, the programs often do not provide the same features, and new features cannot be used with the AOA format. The virtual abandonment of the AOA format by the software industry probably signals its eventual demise.

Figures 3.17 and 3.18 show the results of a computer-generated analysis of the network of Fig. 3.16. Figure 3.17 is a computer sort of the schedule activities by activity identification number. Each activity in a CPM schedule is assigned a unique identification number, which the computer uses to identify the activity. It should be noted that a computer is not required to analyze a CPM schedule: It can be done by hand. For most schedules, however, the computer is quicker and more accurate. Figure 3.18 is a total float sort, in which the computer has listed the activities in order, from least to most float.

Some of the terms used in CPM schedules are as follows:

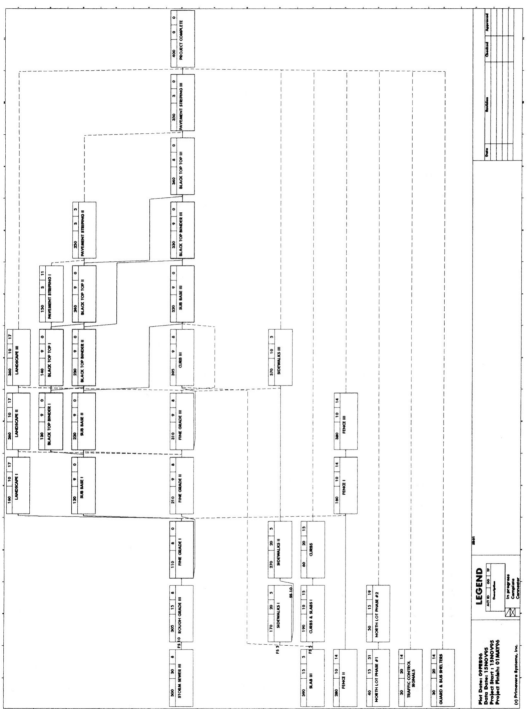

Figure 3.16 Pure logic diagram for an airport employee parking lot.

```
                                    PRIMAVERA PROJECT PLANNER              AIRPORT EMPLOYEE PARKING LOT

REPORT DATE  9FEB96  RUN NO.  70                                           START DATE 15NOV95  FIN DATE  1MAY96
             8:20
ACTIVITY ID REPORT                                                         DATA DATE  15NOV95  PAGE NO.    1
```

ACTIVITY ID	ORIG DUR	REM DUR	%	ACTIVITY DESCRIPTION	EARLY START	EARLY FINISH	LATE START	LATE FINISH	TOTAL FLOAT
20	20	20	0	TRAFFIC CONTROL SIGNALS	15MAR96*	11APR96	4APR96	1MAY96	14
30	20	20	0	GUARD & BUS SHELTERS	15MAR96*	11APR96	4APR96	1MAY96	14
40	15	15	0	NORTH LOT PHASE #1	10JAN96*	30JAN96	21MAR96	10APR96	51
50	15	15	0	NORTH LOT PHASE #2	15MAR96*	4APR96	11APR96	1MAY96	19
60	20	20	0	CURBS	14MAR96	10APR96	4APR96	1MAY96	15
110	8	8	0	FINE GRADE I	15FEB96*	26FEB96	15FEB96	26FEB96	0
120	9	9	0	SUB BASE I	27FEB96	8MAR96	27FEB96	8MAR96	0
130	9	9	0	BLACK TOP BINDER I	11MAR96	21MAR96	11MAR96	21MAR96	0
140	9	9	0	BLACK TOP TOP I	22MAR96	3APR96	22MAR96	3APR96	0
150	3	3	0	PAVEMENT STRIPING I	4APR96	8APR96	19APR96	23APR96	11
160	10	10	0	LANDSCAPE I	27FEB96	11MAR96	21MAR96	3APR96	17
170	20	20	0	SIDEWALKS I	29FEB96	27MAR96	7MAR96	3APR96	5
180	10	10	0	FENCE I	15MAR96	28MAR96	4APR96	17APR96	14
190	10	10	0	CURBS & SLABS I	29FEB96	13MAR96	21MAR96	3APR96	15
210	9	9	0	FINE GRADE II	15FEB96	27FEB96	27FEB96	8MAR96	8
220	9	9	0	SUB BASE II	11MAR96	21MAR96	11MAR96	21MAR96	0
230	9	9	0	BLACK TOP BINDER II	22MAR96	3APR96	22MAR96	3APR96	0
240	9	9	0	BLACK TOP TOP II	4APR96	16APR96	4APR96	16APR96	0
250	3	3	0	PAVEMENT STRIPING II	17APR96	19APR96	24APR96	26APR96	5
260	10	10	0	LANDSCAPE II	12MAR96	25MAR96	4APR96	17APR96	17
270	20	20	0	SIDEWALKS II	14MAR96	10APR96	21MAR96	17APR96	5
280	10	10	0	FENCE II	1MAR96*	14MAR96	21MAR96	3APR96	14
300	30	30	0	STORM SEWER III	15NOV95	28DEC95	28NOV95	10JAN96	8
305	15	15	0	ROUGH GRADE III	15JAN96	2FEB96	25JAN96	14FEB96	8
310	9	9	0	FINE GRADE III	28FEB96	11MAR96	11MAR96	21MAR96	8
320	9	9	0	SUB BASE III	22MAR96	3APR96	22MAR96	3APR96	0
330	9	9	0	BLACK TOP BINDER III	4APR96	16APR96	4APR96	16APR96	0
340	8	8	0	BLACK TOP TOP III	17APR96	26APR96	17APR96	26APR96	0
350	3	3	0	PAVEMENT STRIPING III	29APR96	1MAY96	29APR96	1MAY96	0
360	10	10	0	LANDSCAPE III	26MAR96	8APR96	18APR96	1MAY96	17
370	10	10	0	SIDEWALKS III	11APR96	24APR96	18APR96	1MAY96	5
380	10	10	0	FENCE III	29MAR96	11APR96	18APR96	1MAY96	14
390	15	15	0	SLAB III	1FEB96*	21FEB96	8FEB96	28FEB96	5
395	9	9	0	CURB III	28FEB96	11MAR96	11MAR96	21MAR96	8
400	0	0	0	PROJECT COMPLETE	2MAY96	1MAY96	2MAY96	1MAY96	0

Figure 3.17 Activity identification report.

Report date—the date the computer sort was prepared.

Start date—the project's anticipated start date.

Fin date—the project's scheduled finish date.

Data date—the date from which the schedule is calculated. For the initial schedule, the data date and the start date are usually the same.

Activity ID—the activity's identification number.

Orig dur—the original duration of each activity. Durations are usually calculated in work days. The computer converts work days to calendar days by adding weekends and holidays as directed by the scheduler.

Rem dur—the remaining duration of each activity. In the initial project schedule, the original duration and remaining duration of each activity are

```
                          PRIMAVERA PROJECT PLANNER              AIRPORT EMPLOYEE PARKING LOT

REPORT DATE  9FEB96  RUN NO.  69                                 START DATE 15NOV95  FIN DATE 1MAY96
             8:20
TOTAL FLOAT/EARLY START REPORT                                   DATA DATE  15NOV95  PAGE NO.   1
```

ACTIVITY ID	ORIG DUR	REM DUR	%	ACTIVITY DESCRIPTION	EARLY START	EARLY FINISH	LATE START	LATE FINISH	TOTAL FLOAT
110	8	8	0	FINE GRADE I	15FEB96*	26FEB96	15FEB96	26FEB96	0
120	9	9	0	SUB BASE I	27FEB96	8MAR96	27FEB96	8MAR96	0
130	9	9	0	BLACK TOP BINDER I	11MAR96	21MAR96	11MAR96	21MAR96	0
220	9	9	0	SUB BASE II	11MAR96	21MAR96	11MAR96	21MAR96	0
140	9	9	0	BLACK TOP TOP I	22MAR96	3APR96	22MAR96	3APR96	0
230	9	9	0	BLACK TOP BINDER II	22MAR96	3APR96	22MAR96	3APR96	0
320	9	9	0	SUB BASE III	22MAR96	3APR96	22MAR96	3APR96	0
240	9	9	0	BLACK TOP TOP II	4APR96	16APR96	4APR96	16APR96	0
330	9	9	0	BLACK TOP BINDER III	4APR96	16APR96	4APR96	16APR96	0
340	8	8	0	BLACK TOP TOP III	17APR96	26APR96	17APR96	26APR96	0
350	3	3	0	PAVEMENT STRIPING III	29APR96	1MAY96	29APR96	1MAY96	0
400	0	0	0	PROJECT COMPLETE	2MAY96	1MAY96	2MAY96	1MAY96	0
390	15	15	0	SLAB III	1FEB96*	21FEB96	8FEB96	28FEB96	5
170	20	20	0	SIDEWALKS I	29FEB96	27MAR96	7MAR96	3APR96	5
270	20	20	0	SIDEWALKS II	14MAR96	10APR96	21MAR96	17APR96	5
370	10	10	0	SIDEWALKS III	11APR96	24APR96	18APR96	1MAY96	5
250	3	3	0	PAVEMENT STRIPING II	17APR96	19APR96	24APR96	26APR96	5
300	30	30	0	STORM SEWER III	15NOV95	28DEC95	28NOV95	10JAN96	8
305	15	15	0	ROUGH GRADE III	15JAN96	2FEB96	25JAN96	14FEB96	8
210	9	9	0	FINE GRADE II	15FEB96	27FEB96	27FEB96	8MAR96	8
310	9	9	0	FINE GRADE III	28FEB96	11MAR96	11MAR96	21MAR96	8
395	9	9	0	CURB III	28FEB96	11MAR96	11MAR96	21MAR96	8
150	3	3	0	PAVEMENT STRIPING I	4APR96	8APR96	19APR96	23APR96	11
280	10	10	0	FENCE II	1MAR96*	14MAR96	21MAR96	3APR96	14
20	20	20	0	TRAFFIC CONTROL SIGNALS	15MAR96*	11APR96	4APR96	1MAY96	14
30	20	20	0	GUARD & BUS SHELTERS	15MAR96*	11APR96	4APR96	1MAY96	14
180	10	10	0	FENCE I	15MAR96	28MAR96	4APR96	17APR96	14
380	10	10	0	FENCE III	29MAR96	11APR96	18APR96	1MAY96	14
190	10	10	0	CURBS & SLABS I	29FEB96	13MAR96	21MAR96	3APR96	15
60	20	20	0	CURBS	14MAR96	10APR96	21MAR96	1MAY96	15
160	10	10	0	LANDSCAPE I	27FEB96	11MAR96	21MAR96	3APR96	17
260	10	10	0	LANDSCAPE II	12MAR96	25MAR96	4APR96	17APR96	17
360	10	10	0	LANDSCAPE III	26MAR96	8APR96	18APR96	1MAY96	17
50	15	15	0	NORTH LOT PHASE #2	15MAR96*	4APR96	11APR96	1MAY96	19
40	15	15	0	NORTH LOT PHASE #1	10JAN96*	30JAN96	21MAR96	10APR96	51

Figure 3.18 Total float sort.

the same. (The initial project schedule, once submitted and approved by the owner, is sometimes called the *as-planned schedule*.)

Percent (%)—the percentage of completion of each activity. In the initial schedule, this number should be zero for every activity.

Activity description—a brief description in words of the work represented by each activity. Each activity description should clearly represent the work contained within the activity. Avoid using ambiguous or undefined descriptions such as "start paving" or "finish excavation." The terms "start" and "finish" cannot be measured, and thus do not provide a clear description of the work represented by the activity.

Early start—the earliest date on which the activity could start and still finish the work that logically preceded it.

Early finish—the earliest date on which the activity could finish if it starts on its early start date.

Late start—the latest date on which an activity could start and still finish by the overall project completion date. If an activity starts later than its late start date, the project will be delayed and the schedule will show a later project completion date.

Late finish—the last date on which an activity can finish and still finish the project by the scheduled date.

Total float—a number, expressed usually in work days, that indicates the flexibility available to the project manager for scheduling the performance of the work. Mathematically, total float is the difference between the early and late dates. In Fig. 3.17, for example, activity no. 250, Pavement Striping II, has an early start date of April 17, 1996, and a late start date of April 24, 1996. The difference is 7 calendar days, which converts to 5 work days. As indicated in the total float column, activity no. 250 has 5 work days of float. Float is sometimes called *slack time*. Essentially, float is a measure of how long one may wait before a work activity must be started or how long an activity's duration may be extended or interrupted before the project completion date is delayed. It is also a measure of how much work can be added to a path before the project completion date is affected.

The project schedule is usually developed by the general contractor or the construction manager. Even when the owner runs the schedule on the computer, the data are usually provided by the contractor. This makes sense, since in large part the schedule is a reflection of how the contractor plans to perform the work. Thus, activity durations are a function of the size and experience of the crews to be used and the equipment the contractor plans to mobilize. The logic may also reflect the number of crews the contractor plans to mobilize. For example, will all pile-driving work be done sequentially, or will the contractor mobilize two pile-driving crews and perform some of the pile-driving work concurrently?

The duration of an activity should not be greater than 15 work days or less than 1 work day. The high limit helps assure that the schedule is prepared with adequate detail. The low limit helps assure that the schedule is not overly detailed and also represents the limit of some scheduling programs. Some programs, however, are capable of scheduling a project in increments of 1 hour.

The *critical path* is the path of least float through a project. If the project completion date or intermediate milestones are not fixed when the computer analysis is performed, the critical path of work will have zero float. Zero float means that the late and early dates are the same and the activity must be started and finished on these dates for the project to be completed on time. If the project completion date is fixed in the analysis, then the critical path may have positive, negative, or zero float. Zero float means that the project is expected to finish by the fixed dates, but no earlier or later. Positive float on the critical path means that the project can be finished earlier than the fixed dates. Negative float means that the project cannot be completed as scheduled by the fixed dates. When the critical or any other path in the schedule has negative float, the late dates will be earlier than the early dates. The magnitude of the negative float on the critical path is sometimes used as a rough measure of project delay.

There are several approaches to the development of a CPM schedule. To a certain extent, the approach used depends on the knowledge and experience of the parties developing the schedule. Keep in mind that a CPM schedule can be used to schedule the design process and the owner's overall program.

While such a schedule is almost always a good idea, on fast-tracked projects it is essential, because the design and construction process are integrated on a fast-tracked project. Figure 3.10 showed a simplified example of a fast-tracked project schedule. A schedule integrating both design and construction is typically prepared by the owner. If the owner does not have the expertise to prepare such a schedule, then this task often falls to the project's program manager or designer/builder.

CPM schedule development begins with the identification of all the work that will be required to complete the project. This includes construction work, the procurement process, permitting, and other activities that have an impact on the timing of the project. Contract requirements, such as the time the owner will require to process submittals, should also be considered. Each participant in the project must indicate the amount of time that will be needed to perform its work.

In addition to all work activities and the time required to perform these activities, the relationship among these activities must be identified. With this information in hand, a network diagram can be prepared. Some programs allow the network to be drawn directly on the computer screen. These programs may also allow project schedule information to be entered from this screen. Once the schedule has been developed and analyzed, the next step is to review it to make sure it makes sense, is free of bugs, and conforms to the contract requirements.

It is not enough to prepare one initial project schedule. A schedule is a plan, a best guess. It may be an educated guess, based on years of experience, but it cannot anticipate every event. As events occur that change the plan, so must the schedule change. In addition, with the passing of each work day, work on the schedule is accomplished, changes are made, and work is added to the schedule. The schedule should be modified to reflect the completion of work activities. It should also be modified to reflect the addition or deletion of work. To keep the schedule current, these modifications must be incorporated. Modifications are incorporated when the schedule is updated or "statused." A schedule update is commonly prepared monthly. It is prepared by advancing the data date to the date of the update. Actual start dates are provided for activities that have started. Actual finish dates are provided for activities that have been finished. For activities that have started but not finished, the remaining duration of the activity or the percentage of completion should be indicated (there is no need to provide both, as the computer will calculate whichever is not provided). If changes have been made, then activities representing the change should be added to the schedule, or the existing activities modified or deleted. With the update information inserted and the schedule run again, new early and late dates will be calculated for each activity and a new project completion date projected. If this date is later than the date in the original schedule or previous update, then the project has been delayed during the update period. Though the contract may not require it, it is a good idea to maintain a written record of all revisions that are made to the schedule each month. This should include listing all activities added, deleted, or modified. Some contracts require that a narrative report actually be prepared to accompany the schedule, describe all modifications made, and the reason for any delay or improvement in the project completion date and intermediate milestones. An example of a narrative report is given in Fig. 3.19.

Schedule updates are very useful on projects where contamination is discovered, since any modifications or slow progress that results from the discovery will be recorded in the schedule.

Communication. Though the owner often has the ultimate responsibility to respond to problems, no response is possible until the owner is aware of the problem. Thus, when contamination is encountered on a project it is very often the contractor who must communicate the problem to the project team. In fact, this burden is often placed on the contractor by the contract in the form of a notice provision. The contractor's interest in and obligation to communicate goes beyond mere notice, however. In order for the owner and design team to understand the importance and priority of the discovery, the contractor must also communicate the probable impact of the discovery and the consequences of losing further time. Thus, the contractor must not only communicate the discovery of contamination, but the likely consequences of this discovery from the contractor's perspective. This information can then be used by the owner to mobilize the forces necessary to address the discovery.

The following is a good example of a detailed contract notice provision. It exemplifies the kind of information that can be useful when a contractor is providing notice to the owner of the discovery of hazardous materials.

104.09 Notification of Changes. The primary purpose of this Subsection is to obtain prompt reporting of state conduct or other occurrences which the contractor believes to constitute a change to this contract. Except for changes identified as such pursuant to subsections 104.02 and 104.03, the contractor shall promptly notify the engineer in writing, on forms provided by the Department, and in any event within 5 days from the date that the contractor identifies any state conduct or other occurrence including actions, inactions, and written or oral communications, which the contractor regards as a change to the contract terms and conditions. In no event will the contractor begin work nor incur any expenses with relation to the claimed change prior to giving notice.

The notice shall state, on the basis of the most accurate information available to the contractor:

> **Narrative Report**
>
> **January 31, 1996**
>
> I. Progress
>
> Concrete
>
> Snow and generally poor weather caused us to stop work on site for a period of 2 weeks. This stoppage has caused the completion of list 2 on the south side to slide to February 12th +/−. This slippage has not affected the end date, as we had not originally envisioned the startup of the next lift (lift 3) until March 4 +/−.
>
> At this time, it appears that we will partially start lift 3 sooner, as we have the resources available. (Note: The lift 3 start on the 01/23/96 plot shows a start date of 03/11/96, which is incorrect. We will correct it on the next plot.)
>
> Because of the slippage, we do not expect to start the misc. metals on the north side or install the submarine cable brackets. These two activities were advanced to take advantage of available equipment in the February shutdown but are not essential until much later and have been rescheduled to these later dates.
>
> Steel Fabrication
>
> The steel fabricator has slipped on its delivery date, which now indicates a negative float of 17 days. They are attempting to recover this time and we are still basing our erection around a 07/05/96 delivery.
>
> Machinery Fabrication
>
> Machine fabrication has indicated they will need more time to complete if painting is included in their scope. We are seeking alternative plans.
>
> The gear supplier is behind schedule (4 weeks +/−) on the reducers and is seeking corrective measures. The release of 1 unit not included in the testing phase will facilitate this. They are also behind schedule (8 weeks +/−) on the air buffer manufacturing. We are researching corrective measures but are not confident of their ability to catch up.
>
> II. Revisions
>
> The duration of the erection of the north crane trestle was increased to 2 weeks. The start date of the north pier lift 4 concrete was moved from March 18 to March 4.
>
> We added two blocks of work previously not called out on the master schedule. They are the south and north pit stair towers, which will follow the south lift 4.
>
> The service building has been accelerated to start as soon as the north pier forms are available in an attempt to build in some float and avoid a delay in the start of the steel erection. This item will run parallel with lift 4 on the south side and precede the north side operator's house as it utilizes the same forms.
>
> III. Other
>
> We are presently entering the detailed steel schedule we reviewed on 01/25/96 into a fragnet. We would like to replace the existing steel erection activities with a simplified overall steel schedule on the master schedule backed up by the detailed fragnet at our update, with your permission.

Figure 3.19 Narrative report.

(a) the date, nature, and circumstances of the conduct or occurrence regarded as a change;
(b) the name, function, and activity of each State individual and official or employee involved in or knowledgeable about such conduct or occurrence;
(c) the identification of any documents and the substance of any oral communication involved in such conduct or occurrence;
(d) in the instance of alleged acceleration, the scheduled performance or delivery;
(e) in the instance of alleged extra work, the basis for the contractor's claim that the work is extra;

(f) the particular elements of contract performance for which the contractor may seek additional compensation under this Section 104 including:
  (1) what Pay Item(s) have been or may be affected by the alleged change;
  (2) what labor or materials or both have been or may be added, deleted, or wasted by the alleged change and equipment idled, added, or required for additional time;
  (3) to the extent practicable, what delay and disruption in the manner and sequence of performance and effect on continued performance have been or may be caused by the alleged change;
  (4) what adjustments to contract price, delivery schedule, and other provisions affected by the alleged change are estimated; and
  (5) the contractor's estimate of the time within which the State must respond to the contractor's notice to minimize cost, delay, or disruption of performance.

Following submission of the notice, the contractor shall diligently continue performance of this contract to the maximum extent possible in accordance with the contract documents, unless such notice results in a direction of the engineer or a communication from an authorized representative of the engineer, in either of which events the contractor shall continue performance in compliance therewith, provided, however, that if the contractor regards such direction or communication itself as a change, notice shall be given as provided above. All directions, communications, interpretations, orders, and similar actions shall be reduced to writing promptly and copies thereof furnished to the contractor and the engineer. The engineer shall promptly countermand any action which exceeds the authority of the authorized representative.

The failure of the contractor to give notice pursuant to the provisions of this subsection shall constitute an absolute waiver of any and all claims which may arise as a result of the alleged change. Moreover, no action or inaction of any person shall constitute a waiver of the state's absolute right to receive written notice pursuant to this subsection of an alleged claim.

Summary

Each party to the contract plays a role in preparing for the discovery of contamination on a project. These roles are not special or peculiar, but extensions of the traditional roles of the owner, designer, and the contractor on the site, with an emphasis on planning for and addressing the discovery of contamination on the project site.

Chapter 4

The Contract as a Risk Management Tool

Among its many uses, a construction contract apportions risk. A number of standard contract forms are available. Almost every professional and industry organization has produced a standard form. The most common of these forms are those sold by the American Institute of Architects (AIA). Standard contract forms, however, are also produced by the Associated General Contractors (AGC), the National Society of Professional Engineers (NSPE), and the Construction Managers Association of America (CMAA), to name a few. In addition, several owners have produced their own standard contract documents and clauses. For example, the federal government has a number of standard clauses which it has compiled in the *Federal Acquisition Regulations* (FAR). Other organizations using standard contract provisions include state departments of transportation, school systems, and other organizations that do a lot of construction. Whenever a standard contract form is used, keep in mind that it is a *standard* form. It was not designed for any particular project. Standard forms are intentionally generic so that they can be used for the broadest possible range of projects and by the greatest number of owners, designers, contractors, subcontractors, and suppliers. To be used on any particular project, they must be supplemented, modified, or otherwise altered. Sometimes the standard documents themselves facilitate tailoring. For example, the AIA documents come in many forms. Standard AIA contracts are available for agreements between architects and owners, owners and contractors, owners and construction managers, contractors and subcontractors, etc. The standard AIA forms cover the range of contract types from lump sum to cost plus, with a fixed fee, and with shared savings. Each AIA form provides for some customization. Blanks are provided for such things as the names of the contracting parties, dates, and contract amounts. In certain instances, options are provided. Beyond these basic modifications, however, the AIA forms have a few additional customizing features. AIA forms, like any standard contract forms, do not have to be used precisely as prepared. In fact, in many cases these standard contract forms are modified quite dramatically. Do not hesitate to add, subtract, or modify the contract clauses for standard contract forms to tailor the standard forms to a particular project. When modifying standard forms, however, or when writing contracts from scratch, keep in mind the following guidelines.

1. Modifying one part of a standard contract form may affect another part. When customizing a contract, read the entire contract to be sure that the changes made do not alter the meaning of other contract provisions.
2. Make sure that the provisions that are retained in a standard contract actually reflect the parties' intentions. For example, the standard AIA contract includes a provision requiring binding arbitration of disputes. If binding arbitration is not the means by which the parties intend to resolve disputes, then this provision should be stricken and replaced with a clause describing the forum in which the parties will resolve disputes.
3. Do not attempt to revise standard contracts without the assistance or review of legal counsel experienced in the drafting of construction contracts.

With regard to the risk of discovering contamination unexpectedly, several common construction contract provisions are of interest. The first and most important of these clauses is the *changes clause*. The changes clause appears in some form in all construction contracts. In fact, it is the changes clause that makes a construction contract different from other kinds of contracts. A changes clause makes it possible for the parties, particularly the owner, to modify the terms and requirements of the contract without having to renegotiate the entire contract. Few other types of contracts have similar clauses.

The Changes Clause

A good changes clause is generally composed of the following elements:

1. A definition of what constitutes a change
2. A notice provision
3. A description of the procedure by which a change is made to the contract
4. A provision requiring the contractor to perform the changed work
5. A mechanism for payment

Definition of a change

As its name infers, a change is a modification of the contract's requirements. For example, adding asbestos abatement work to a contractor's work, on a project where such work was not originally required, is a change. Changes can have many sources. The owner of the project may make a change to add new work or to delete work that had originally been included. A change may be necessary because of a flaw in the project's design. A contractor may request a change because the contractor believes the actual project work is different from the work described in the contract documents. A well-written changes clause accounts for each of these possibilities.

In some instances, construction contracts anticipate the occurrence of contamination through specific language in the changes clause. In most contracts, however, contamination problems are handled under a clause that is similar to the changes clause and that is called a differing site conditions clause or a concealed conditions clause. Government contracts, particularly federal contracts, typically use the term *differing site condition*. The standard contract form published by the American Institute of Architects refers to *concealed conditions*. Whatever term is used, these clauses generally function the same way, though the federal clause is more specific and more demanding.

The differing site conditions clause arose from federal contracting in the dredging industry. At one time, only the changes clause existed to handle any variations in the project work. Because dredging had so many uncertainties associated with the site, the differing site conditions clause evolved. Not too long ago it was commonly referred to as a *changed conditions* clause. The clause is not significantly different from the standard changes clause, but it focuses on conditions associated with the project site. The following is a typical example of a differing site conditions clause from a government contract.

Differing Site Conditions

During the progress of the work, if preexisting subsurface or latent physical conditions are encountered at the site, differing materially from those indicated in the contract, or if preexisting unknown physical conditions of an unusual nature, differing materially from those ordinarily encountered and generally recognized as inherent in the work provided for in the contract, are encountered at the site, the party discovering such conditions shall promptly notify the other party in writing of the specified differing conditions before they are disturbed and before the affected work is performed.

Upon written notification, the Engineer will investigate the conditions and if the Engineer determines that the conditions materially differ and cause an increase or decrease in the cost or time required for the performance of any work under the contract, an adjustment, excluding loss of anticipated profits, will be made and the contract modified in writing accordingly. The Engineer will notify the Contractor in writing of the Engineer's determination whether or not an adjustment of the contract is warranted.

No contract adjustment which results in a benefit to the Contractor will be allowed unless the Contractor has provided the required notice.

The adjustment will be by agreement with the Contractor. If the parties are unable to agree, the Engineer will determine the amount of the equitable adjustment by establishing the costs or by using force account, and adjust the time as the Engineer deems appropriate. Extensions of time will be evaluated in accordance with Section 1-05.1.

Because the differing site conditions clause is central to defining the risk to the contracting parties of the unexpected discovery of contamination, it is worthy of additional discussion. First, this clause specifically addresses situations where something is discovered on the project site that is different from what was described in the contract documents or that might otherwise be expected. Thus, it specifically addresses the unexpected discovery of contamination, either because the contamination was found at a location different from the location identified in the plans or was more extensive than shown, because no contamination was expected, or because contamination was found in an unusual place. Consequently, the parties' contractual rights and obligations regarding the unexpected discovery of contamination are most commonly defined by this clause if it is included in the contract. A considerable body of legal precedent relates to the meaning and interpretation of the terminology used in differing site conditions clauses, particularly the version of the clause found in federal contracts.

There are two types of differing site conditions. The first type is called a type 1 differing site condition and is characterized as subsurface or latent physical conditions at the site that differ materially from those indicated in the contract. This category of changed conditions is not restricted to subsurface conditions, but includes conditions which may be at or above the surface that are latent in the sense that they are concealed, hidden, or dormant.

Proof of the existence of a type 1 differing site condition depends on a comparison of the actual conditions encountered with those indicated in the contract. If the contract contains no indications with respect to subsurface conditions, a type 1 differing site condition cannot be encountered, since the actual conditions encountered cannot be described as differing from those indicated in the contract. Relief, however, may still be available under these circumstances within the second category of changed conditions covered by the clause.

Examples of type 1 differing site conditions include:

1. The presence of contaminants such as lead, asbestos, or PCBs, when the contract documents indicate that abatement, remediation, or other cleanup activities have occurred
2. The presence of rock or boulders in an excavation area where none are shown or indicated, or the existence of such rock at materially different elevations than was indicated in the data available to bidders
3. The presence of subsurface water where none is indicated by the contract documents
4. The failure of designated borrow pits or quarry sites to produce the required materials entirely or in sufficient quantities, without excessive waste of unusable materials beyond that reasonably anticipated from the prebid data
5. The presence of rock, debris, or other subsurface obstructions in substantially greater quantities than shown in the contract documents
6. The existence of a subfloor not shown on the drawings, which must be removed in order to renovate a building under the contract

7. The encountering of groundwater at a higher elevation, or in quantities in excess of those indicated or reasonably anticipated from the data available to bidders

8. The encountering of groundwater contamination at greater concentrations than indicated in the contract documents

9. The encountering of rock that is materially harder or tougher to excavate or drill and blast than was expected from information available prior to bidding

10. The presence of a higher moisture content in soils to be compacted than was anticipated from the contract data

11. Ground contour elevations at the site which differ from those shown on the drawings, and which accordingly require greater quantities of excavation or fill

The existence of type 1 differing site conditions is dependent on a comparison between the conditions actually encountered and those described in the contract. The contract indications as to hidden or latent conditions need not be explicit or specific; all that is required is that there is a sufficient indication to support the conclusion that a bidder would not reasonably have expected the conditions actually encountered.

The second type of differing site condition is known as a type 2 differing site condition and is described as encountering a previously unknown physical condition or conditions at the site which is (are) of an unusual nature or differ materially from those ordinarily encountered and generally recognized as inherent in work of the character provided for in the contract.

This category of differing site condition is not predicated on the existence of some difference between the conditions encountered and those represented or indicated in the contract. Instead, the conditions encountered must be unknown, unusual, and differing materially from what ordinarily would be expected in performing the kind of work called for in the contract. Accordingly, the standard applied is the *standard of normal conditions*.

Examples of conditions which have been recognized as providing the basis for relief as a type 2 differing site condition include the following:

1. The unexpected and highly corrosive nature of groundwater at the site, which resulted in extensive damage to the contractor's dewatering equipment and extensive cleanup efforts

2. Excessive hydrostatic pressure encountered in the laying of a pipeline, which could not have been anticipated at the time of bid

3. Jet fuel which flooded manholes because of an unknown blockage in an airport drainage system which caused damage to underground transmission cables installed by the contractor, and which necessitated extensive cleanup efforts

4. The presence of caked material which was not revealed by a site inspection of heating ducts to be cleaned under the contract

5. An unknown and unanticipated oily substance which prevented adherence of polyvinyl chloride that the contract required be applied to the roof, which also required extensive cleanup efforts

6. Failure of rock from an approved borrow pit to fracture in the manner expected for production of contract aggregate

The unknown and unusual condition encountered does not have to be a complete freak, such as permafrost in the tropics. All that is required is that the condition must be one that was reasonably unanticipated based on the contract documents. In demonstrating the existence of a type 2 differing site condition, the task is not as clear cut as in a type 1 situation. In the type 2 situation the analyst must compare the actual conditions encountered with what one would reasonably expect, taking into account all those factors that an accomplished bidder customarily uses regarding quantity, quality, and methods of performing the particular work.

The customary elements in establishing a type 2 differing site condition are based on the bidding judgments that were employed, including a reasonable site inspection, a study of the contract documents, and intelligent interpretation based on construction experience. Some of the factors that form the background for establishing that the condition was not reasonably to be anticipated are the customs of the trade, common knowledge of the industry, manufacturer instructions, recommendations, and traditional assumptions involved in bidding a particular kind of work.

Whether the conditions discovered represent a type 1 or a type 2 differing site condition, the next essential step in the proper management of the project and the protection of the rights provided by the differing site conditions clause is to provide proper notice.

Notice

Notice is an important component of any changes clause, whether it is a general changes clause or a more specific clause such as a differing site conditions clause. A notice provision requires that the party to the contract who discovers the change (the differing site condition) tell the other party to the contract of the problem. For example, if the contractor discovers an abandoned fuel tank that still contains fuel, the contractor must notify the other contracting entity, usually the owner, of the discovery. In this instance, notification is required to afford the owner an opportunity to investigate the problem and come up with the most economical fix. Failing to provide notice before proceeding to drain and remove the tank denies the owner the ability to consider other, more economical alternatives. In legal terms, failing to provide adequate notice may prejudice the owner's ability to mitigate costs or other damages. The changes clause deals with this problem simply. Failure to provide notice results in the contractor losing the right to added compensation or a time extension for costs and delays resulting from draining and removing the tanks or otherwise remedying the problems associated with the differing site condition or change.

Many changes clauses require notice in writing within a certain period after the time of discovery. For a differing site condition the time to provide notice is essentially immediate, and certainly before the conditions are disturbed. Though the courts and boards in some jurisdictions do not always enforce the notice provisions of the changes clause, the parties to a contract ignore such provisions at their peril. For example, though a contractor may not have provided formal written notice, if the contractor informed the owner's representative orally and the owner's representative recorded this oral notice in writing in his or her daily log, some jurisdictions have deemed that notice was provided. Similarly, if notice was given in a project meeting and recorded in the minutes of the meeting, notice again may be deemed to have been provided.

A typical notice provision was provided and discussed in Chap. 3.

Procedures for modifying the contract

The third component of a good changes clause is a provision that describes how the contract is to be modified in the event of a change. Typically the procedure will involve acknowledgment by the parties to the contract that a change exists. This is usually followed by a requirement that the contractor prepare a proposal. The preparation of a proposal is dependent on the directions given by the owner. The contractor must have clear guidelines as to what is required in order to prepare a proposal. It is not good practice to require the contractor to propose a solution to the change, though clearly the contractor is not prevented from doing so. Developing the scope of changed work, however, is generally the owner's responsibility. In its proposal, the contractor presents the estimated costs and time associated with performing the changed work. Upon receipt of the proposal, the parties then sit down at the negotiating table and hammer out an acceptable price for the new work. They also determine whether the contract completion date needs to be extended to accommodate the changed work. Once agreement is reached, the contract documents are modified in writing. This written modification is often called a *change order*. Every change order should address both the added cost of the change and any extension of the contract completion date. If no extension of time is necessary, the change order should say so. It is important that the change order make clear that all aspects of the change and its costs were discussed, negotiated, and are addressed by the time extension and added compensation provided by the change. The following is language that might appear above the contractor's signature on a change order.

Contractor's Acceptance

The terms and conditions of this extra work order, including the amount and time contained herein, constitute a full accord and satisfaction by the administration and the contractor for all costs and time of performance related to the actions described or referenced herein, including, but not limited to, delay and impact costs resulting from this extra work order.

Except as amended herein, all provisions of said contract remain in full force and effect.

Continuation of work

The contract rarely gives a contractor an option to perform the extra or changed work. If the owner directs a change, then the contractor must perform the changed work. The clause or provision that makes this possible is often called a *continuation-of-work clause*. It is usually part of the changes clause. Typically, such a clause also obligates the surety to cover the added work under the bond, if a bond is required on the project. A typical continuation-of-work clause reads as follows:

> In the event that the contractor and owner cannot agree upon the value of the change, at the direction of the owner, the contractor shall commence the change work. All records of costs expended will be kept by the owner and agreed upon on a daily basis.

The only exception to the continuation-of-work clause is a change made by the owner that could be characterized as a cardinal change. A *cardinal change* is a change made by the owner that is outside the scope of the work contemplated. Though the concept is rarely used to escape the performance of work changed by the owner, the concept of a cardinal change potentially has great applicability to a situation where contamination is discovered unexpectedly. Hazardous waste remediation is often far beyond the capabilities of many contractors and often completely outside the scope of work contemplated by the parties to a contract. Even if a contractor wanted to tackle the cleanup of contamination unexpectedly discovered on the site, licensing and insurance requirements might bar it from performing such work. Consequently, directing a contractor to clean up contamination on a project where the discovery of the contamination was unexpected could be a valid example of a cardinal change. In such cases, proceeding with the work could be at the contractor's option. The issue is one of degree. Certainly a contractor could not refuse to clean up a small fuel spill, particularly if the contractor was the source of the spill. One rule of thumb to consider when evaluating whether or not a change is a cardinal change is that if the work described by the change could reasonably be accomplished by a separate contract, then it is possible the change is a cardinal change. Keep in mind that refusing to proceed with changed work directed by the owner when the work is not a cardinal change might be considered a breach of contract and cause for termination for default. Thus, refusing to proceed with an owner's directive is not an action that should be taken without careful consideration and consultation with qualified legal counsel.

Methods of payment

A good changes clause should also define how a contractor will be paid for the change. Almost all changes clauses provide for three methods of pricing the work. These three methods are negotiated price, unit prices established by the contract, or time and materials payment. Most likely, no unit price will be established for the occurrence of contamination. Therefore, one of the other two methods must be used. In instances where a price cannot be agreed to or the extra work required cannot be quantified, the contract may allow the contractor to proceed with the work on a time-and-materials or force account basis. *Force account* is the term used for payments made on a time-and-materials basis on highway construction jobs. Time-and-materials work is also sometimes called *cost-plus* work, because the contractor is paid its actual costs for labor, material, and equipment, plus some percentage markup for overhead and profit. In a good changes clause, each of these cost components is described in detail, together with a description of the documentation that must be provided to prove that the costs were actually incurred.

The following is a good example of the pricing portion of a changes clause:

109.04 DIFFERING SITE CONDITIONS, CHANGES, EXTRA WORK AND FORCE ACCOUNT WORK. Differing site conditions, changes, and extra work performed under Section 104 will be paid for using the following methods as appropriate:

A. Contract unit prices.
B. Unit prices agreed upon in the change order authorizing the work.
C. A lump-sum amount agreed upon in the change order authorizing the work.
D. If directed by the Department, work performed on a force account basis is to be compensated in the following manner:

1. *Labor.* For all necessary labor and foremen in direct charge of the specific operations, whether the employer is the Contractor, subcontractor or another, the Contractor shall receive the rate of wage (or scale) actually paid as shown in its certified payrolls for each and every hour that said labor and foremen are actually engaged in such work.

   The Contractor shall receive the actual costs paid to, or on behalf of, workmen by reason of health and welfare benefits or other benefits, when such amounts are required by collective bargaining agreements or other employment contracts generally applicable to the classes of labor employed on the Work.

2. *Bond, Insurance, and Tax.* For bond premiums, property damage, liability, and workmen's compensation insurance premiums, unemployment insurance contributions, and social security taxes on the force account work, the Contractor shall receive the actual incremental cost thereof, necessarily and directly resulting from the force account work. The Contractor shall furnish satisfactory evidence of the rate or rates paid for such bond, insurance, and tax.

3. *Materials.* The Department reserves the right to furnish such materials as it deems advisable, and the Contractor shall have no claims for costs and markup on such materials.

   Only materials furnished by the Contractor and necessarily used in the performance of the work will be paid for. The cost of such materials shall be the cost to the purchaser, whether Contractor, subcontractor or other forces from the supplier thereof, together with transportation charges actually paid by him, except as the following are applicable:

    a. If a cash or trade discount by the actual supplier is offered or available to the purchaser, it shall be credited to the State notwithstanding the fact that such discount may not have been taken.

    b. If materials are procured by the purchaser by any method which is not a direct purchase from a direct billing by the actual supplier to such purchaser, the cost of such materials is the price paid to the actual supplier as determined by the Engineer plus the actual costs, if any, incurred in the handling of such materials.

    c. If the materials are obtained from a supply or source owned wholly or in part by the purchaser, the cost of such materials shall not exceed the price paid by the purchaser for similar materials furnished from said source of items or the current wholesale price for such materials delivered to the job site, whichever price is lower.

    d. If the cost of such materials is, in the opinion of the Engineer, excessive, then the cost of such materials is deemed to be the lowest current wholesale price at which such materials are available in the quantities concerned delivered to the jobsite, less any discounts as provided in item a.

    e. If the Contractor does not furnish satisfactory evidence of the cost of such materials from the actual supplier thereof, the cost will be determined in accordance with item d. above.

4. *Equipment and Plant*

    a. *Contractor-Owned Equipment and Plant.* The hourly rates for Contractor-owned equipment and plant will be determined from the applicable volume of the Rental Blue Book (referred to hereafter as the "Blue Book"), published by DATAQUEST, Inc.

    The Blue Book will be used in the following manner:

        (1) The hourly rate will be determined by dividing the monthly rate by 176. The weekly, hourly, and daily rates will not be used.

        (2) The number of hours to be paid will be the number of hours that the equipment or plant is actually used on a specific force account activity.

        (3) The current revisions will be used in establishing rates. The current revision applicable to specific force account work is as of the first day of work performed on that force account work and that rate applies throughout the period the force account work is being performed.

        (4) Area adjustments will be made. Equipment life adjustments will be made in accordance with the rate adjustment tables.

        (5) Overtime shall be charged at the same rate indicated in paragraph (1) above.

        (6) The estimated operating costs per hour will be used for each hour that the equipment or plant is in operation on the force account work. Such costs do not apply to idle time regardless of the cause of the idleness.

        (7) Idle time for equipment will not be paid for, except where the equipment has been held on the project site on a standby basis at

the request of the Engineer and, but for this request, would have left the project site. Such payment will be made at one half the rate established in paragraph (1) above.

(8) The rates established above include the cost of fuel, oil, lubrication, supplies, small tools, necessary attachments, repairs, overhaul and maintenance of any kind, depreciation, storage, overhead, profits, insurance, all costs (including labor and equipment) of moving equipment or plant onto and away from the site, and all incidentals.

(9) Operator costs are not included in the hourly rate for equipment.

All equipment shall, in the opinion of the Engineer, be in good operating condition. Equipment used by the Contractor shall be specifically described and be of suitable size and suitable capacity required for the work to be performed. In the event the Contractor elects to use equipment of a higher rental value than that suitable for the work, payment will be made at the rate applicable to the suitable equipment. The Resident Engineer will determine the suitability of the equipment. If there is a differential in the rate of pay of the operator of oversize or higher rate equipment, the rate paid for the operator is to be that for the suitable equipment.

In the event that a rate is not established in the Blue Book for a particular piece of equipment or plant, the Engineer will establish a rate for that piece of equipment or plant that is consistent with its cost and use in the industry.

The above provisions apply to the equipment and plant owned directly by the Contractor or by entities which are divisions, affiliates, subsidiaries or in any other way related to the Contractor or its parent company.

b. *Rented Equipment and Plant.* In the event that the Contractor does not own a specific type of equipment and must obtain it by rental, the Contractor shall inform the Resident Engineer of the need to rent the equipment and of the rental rate for that equipment prior to using it on the work. The Contractor will be paid the actual rental rate for the equipment for the time that the equipment is actually used to accomplish the work, provided that rate is reasonable, plus the cost of moving the equipment onto and away from the job. The Contractor shall provide a copy of the paid receipt or canceled check for the rental expense incurred.

5. *Profit.* Profit shall be computed at five percent of the following:
   a. Total material cost (bare cost not including FOB).
   b. Total direct labor cost (actual hours worked multiplied by regular hourly rate).

6. *Overhead.* Overhead costs shall be computed at 10 percent of the following:
   a. Total material cost (bare cost not including FOB).
   b. Total direct labor cost (actual hours worked multiplied by regular hourly rate).

7. *Subcontracting.* For administration costs in connection with approved subcontract work, the Contractor shall receive an amount equal to five

percent of the total of such work completed as set forth in item nos. 1 through 6.

8. *Records.* The Contractor shall maintain force account records in such a manner as to provide a clear distinction between the direct costs of work paid for on a force account basis and the costs of other operations.

From the above records, the Contractor shall furnish the Engineer completed daily force account work reports of each day's work to be paid for on a force account basis. Said daily force account work reports shall be signed by the Contractor and submitted daily. The daily force account work reports shall be detailed as follows:

a. Name, classification, date, daily hours, total hours, rate, and extension for each laborer and foreman.
b. Designation, dates, daily hours, total hours, rental rate (including a copy of the Blue Book pages used), and extension or each unit of machinery and equipment.
c. Quantities of materials, prices and extensions.
d. Transportation of materials.
e. Cost of property damage, liability, and workmen's compensation insurance premiums; unemployment insurance contributions; bonds; and social security tax.

Material charges shall be substantiated by valid copies of vendor's invoices. Such invoices shall be submitted with the daily force account work reports, or if not available, they shall be submitted with subsequent daily force account work reports. Should said vendor's invoices not be submitted within 60 days after the date of delivery of the material, or within 15 days after the completion, whichever occurs first, the Department reserves the right to establish the cost of such materials at the lowest current wholesale prices at which said materials are available, in the quantities concerned delivered to the location of work less any discounts provided in subsection 109.04 D3.a.

The Engineer will compare its records with the completed daily force account work reports furnished by the Contractor and make any necessary adjustments. When these daily force account work reports are agreed upon and signed by both parties, said reports become the basis of payment for the work performed, but do not preclude subsequent adjustment based on a later audit by the Department.

The Contractor's cost records pertaining to work paid for on a force account basis shall be open to inspection or audit by representatives of the Department, during the life of the Contract, and for a period of not less than three years after acceptance thereof, and the Contractor shall retain such records for that period. Where payment for materials or labor is based on the cost thereof to forces other than the Contractor, the Contractor shall ensure that the cost records of such other forces are open to inspection and audit by representatives of the Department on the same terms and conditions as the cost records of the Contractor. If an audit is to be commenced more than 60 days after acceptance, the Contractor is provided a reasonable notice of the time when such audit is to begin. In case all or part of such records are not made so available, the Contractor understands and agrees that any terms not supported by reason of such unavailability of the records will not be allowed, or if payment therefore has already been made, the Contractor shall refund to the Department the amount so disallowed.

E. *Actual Costs.* Where agreement on price cannot be reached and work on a force account basis has not been directed by the Department, an adjustment, if required, will be made based on actual costs incurred by the Contractor. Administrative overhead costs will be compensated via a 10% markup for overhead and a 5% markup for profit to be applied to the sum of actual labor, equipment, and material costs. Actual costs submitted to the Department are subject to audit by the Department. The Contractor will provide access to all financial records and cooperate fully with the Department's auditor in the event that the Department decides to audit the Contractor's actual costs.

Note that this clause provides specific markups for overhead and profit. Standard contract forms often fail to address allowable markups for overhead and profit. For example, the standard AIA forms allow "reasonable" markups for overhead and profit. There may be considerable disagreement about what constitutes "reasonable" overhead and profit. It is much better to define these percentages in the contract.

Many changes clauses also mandate that the contractor must have written authorization to proceed with change order work. Often, however, contractors proceed with changed work without written authorization. When this occurs, both the owner paying for the work and the contractor performing the work are taking serious risks. The contractor's risk is that the firm may not be paid for the change order work performed because there was never a written change order. If the owner establishes a practice of paying for change order work performed without written authorization, the owner may not be able to rely on the contract's protections to deny payment for unauthorized work. This is an example of the actions of the parties giving new meaning to the contract.

A well-written changes clause is essential to defining the risks of the parties to the contract with regard to changes or modifications. In the case of the unexpected discovery of contamination, a differing site conditions clause defines the risk with even greater precision. Of particular importance to the functioning of the changes clause is a clear notice requirement, establishing the necessity of timely written notification of the other party to the contract and a penalty for failure to do so. In addition, to facilitate the management and administration of the project, the changes clause should define the procedures by which the contract will be modified and the methods by which payment for the change will be made, and a description of how these amounts will be calculated.

Variation of Quantities

As discussed in Chap. 3, one variation of a fixed-price contract is a unit-price contract. In the unit-price contract, the price that the owner will pay to have a unit of work performed, such as a cubic yard of concrete poured, is fixed, but the quantity is only an estimate. Thus, if the actual quantity of work performed exceeds the estimate, the contractor will be paid for the full actual quantity. If the actual quantity is less than the estimated quantity, the contractor will be paid only for the actual quantity. If the actual quantity is substantially less than the estimated quantity, however, the contractor might not be able to recover the fixed costs of performing the work. For example, if it is necessary to mobilize a crane to remove several buried tanks, but the quantity of tanks actually removed is less than anticipated, then the contractor might not recover even the fixed cost of mobilizing the crane. To address this possibility, a type of clause has been developed called a *variation in quantities clause*. The following is the version of this clause used by the federal government.

> 52.212-11 VARIATION IN ESTIMATED QUANTITY. If the quantity of a unit-priced item in the contract is an estimated quantity and the actual quantity of the unit-priced item varies more than 15 percent above or below the estimated quantity, an equitable adjustment in the contract price shall be made upon demand of either party. The equitable adjustment shall be based upon any increase or decrease in costs due solely to the variation above 115 percent or below 85 percent of the estimated quantity. If the quantity variation is such as to cause an increase in the time necessary for completion, the contractor may request, in writing, an extension of time, to be received by the Contracting Officer within 10 days from the beginning of the delay, or within such further period as may be granted by the Contracting Officer before the date of final settlement of the contract. Upon receipt of a written request for an extension, the Contracting Officer shall ascertain the facts and make an adjustment for extending the completion date as, in the judgement of the Contracting Officer, is justified.

This clause provides for the renegotiation of the unit price in the event that the estimated quantity is significantly in error. Such a clause is useful whenever unit prices are used in a contract. Unit prices are an effective way to deal with the potential for the discovery of contaminants on site, but the quantity cannot be determined. Whenever unit prices are used and appropriate estimated quantities are provided, this clause should be used.

Scheduling

As discussed in Chap. 3, a detailed CPM schedule is recommended for projects where the discovery of contamination is possible. In fact, a good CPM schedule is recommended for almost any project. The following is an example of a good scheduling specification when the contractor is responsible for providing the schedule.

> **108.03 Prosecution of Work**
>
> *Description.* The contractor shall submit to the engineer a schedule of work for approval. The schedule of work shall be used to monitor the sequence of construction operations and the progress of work.
>
> The Schedule of Work shall be in the form specified herein.
>
> *Critical Path Method Schedule.* The construction of this project shall be planned and recorded with a Critical Path Method (CPM) schedule in the form of

an activity on node (AON) diagram. The schedule shall be used for coordinating and monitoring all work under the contract including the activity of subcontractors, vendors, and suppliers.

Preparation of Bid Schedule. Each bidder shall submit a schedule of the work to indicate the scheduled dates or completion of components of the project. It shall include approval, fabrication, and delivery activities precedent to the performance of the major components of the project. The schedule shall contain a minimum of 50 activities. The schedule can be in a barchart format. The schedule must demonstrate the contractor's understanding of the project and ability to:

1. Complete the work by the contract completion date.
2. Meet the milestones specified in Special Provision 108 entitled Control of the Work.

Preparation of 60 Day Schedule. Prior to performing any work on site, the contractor shall provide the engineer a schedule for the first 60 days of work. The schedule can be in a bar chart format. At a minimum, it must show all work activities that require Department approvals prior to beginning the work. The approval activities must be shown on the barchart schedule. The schedule will be used to forecast initial Department inspection resources and submittal review planning.

Preparation of Initial Schedule. Within 30 calendar days of contract award, the contractor shall submit for the engineer's approval a detailed initial schedule. The schedule shall meet the requirements set forth herein.

Within 15 calendar days of the contractor's submittal, the engineer will review the schedule and provide the contractor in writing corrections needed to approve the schedule. The contractor must make all corrections and resolve all comments within 60 calendar days of the Notice to Commence work. If the schedule is not approved within 60 calendar days of the Notice to Commence work, the Department will withhold all contract payments until the schedule is approved.

The approval of the schedule by the Engineer in no way attests to the validity of the assumptions, logic constraints, dependency relationships, resource allocations, manpower and equipment, or any other aspect of the proposed schedule. The contractor is and shall remain solely responsible for the planning and execution of work in order to meet project milestones or contract completion dates.

The construction time for the entire project, or any milestone, shall not exceed the specified contract time. Logic or activity durations will be revised in the event that any milestone or contract completion date is exceeded in the schedule.

Schedule Requirements

1. Activity Information: All AON diagrams shall include:
    a. Activity ID
    b. Activity Description
    c. Finish to Start relationships with no leads or lags
2. Duration (Working Days): No activity will have a duration greater than 15 working days or less than one working day. Activity durations expressed in hours will not be allowed unless approved by the engineer. If requested by the engineer, the contractor shall furnish any information needed to justify the reasonableness of activity time durations. Such information shall include, but not be limited to, estimated activity manpower, unit quantities, and production rates.

3. Procurement and Submittals: Separate procurement into at least two activities—fabrication and delivery. When the procurement also requires a submittal to and approval by the Department, insure these separate activities are shown in the schedule logic. Insure all work activities that require a submittal are preceded by submittal and approval activities.
4. Constraints: Use only contractual constraints in the schedule logic. No other constraints are allowed unless approved by the engineer. The disallowance of constraints includes the use of activity mandatory start and finish dates.
5. Float: Float is defined as the amount of time between when an activity "can start" (the early start) and when an activity "must start" (the late start). It is understood by the Department and the contractor that float is a shared commodity, not for the exclusive use of financial benefit of either party. Either party has the full use of the float until it is depleted.
6. Activity Codes: Activities shall be identified by codes to reflect the following information related to an activity:
    a. Responsible party for the accomplishment of each activity (only one party can be responsible for an activity)
    b. Phase/stage as required by the maintenance and protection of traffic plan and/or the Special Provisions
    c. Area/location
7. Computer Compatibility: The CPM schedule must be processed through a computer and be compatible with Primavera Project Planner Software, Version 5.0 or later, by Primavera Systems, Inc., Bala Cynwyd, Pa. It is the contractor's responsibility to ascertain the software compatibility with the resident engineer.

Initial Schedule Submittal Requirements

1. Predecessor/Successor Sort
2. Total Float/Early Start Sort
3. Responsibility/Early Start Sort
4. Area/Early Start Sort
5. Logic Diagram: Produce diagram with not greater than 100 activities per ANSI D (24-inch×36-inch) size sheet. Insure each sheet includes title, match data or diagram correlation, and a key to identify all components used in the diagram.
6. Narrative discussing general approach to completion of the work
7. Diskette on Primavera (P3) format

Schedule Updates. The contractor shall update the schedule monthly to show current progress. The data date for the update shall be determined by the engineer. The schedule update shall be submitted within seven calendar days of the data date. The engineer may require submission of the updated schedule on diskette prior to submission of the full update package. Should the contractor fail to provide an update, the Department may withhold payment of the current monthly progress estimate until the monthly schedule update is submitted. The update will include:

1. Date of activities' actual start and completion
2. The percent of work remaining for activities started, but not complete as of the update date

3. Narrative report including a listing of monthly progress, the activities that define the critical activities from the previous update, sources of delay, any potential problems, requested logic changes, and work planned for the next month.
4. Predecessor/Successor Sort
5. Total Float/Early Start Sort
6. Responsibility/Early Start Sort
7. Area/Early Start Sort
8. Diskette in Primavera (P3) format
9. Fragnet of logic diagram for all requested logic changes
10. Updated logic diagram as required by the engineer. At a minimum, the Department shall require a final logic diagram at the end of the project showing the planned and actual starts and completions.
11. A barchart comparison of the updated schedule to the initial schedule. This diagram shall show actual and planned performance dates for all completed activities.

Schedule Revisions. The contractor will revise the schedule for the following: a delay in completion of the project or contractual milestones or actual prosecution of the work which is, as determined by the engineer, significantly different than that represented on the schedule: Schedule revisions will be considered incidental to Pay Item 108:51.

Recovery Schedule. If the initial schedule or current updates fail to reflect the project's actual plan or method of operation, or a contract milestone date is more than 30 calendar days behind, the Department may require that a recovery schedule for completion of the remaining contract work be submitted. The Recovery Schedule must be submitted within seven calendar days of the Department's request. The Recovery Schedule shall describe in detail the contractor's plan to complete the remaining contract work by the contract milestone date. The Recovery schedule submittal shall meet the same schedule requirements as the Initial Schedule. The narrative submitted with the Recovery Schedule should describe in detail all changes that have been made to meet the contract milestone date.

Change Orders. When a change order is proposed, the contractor must identify all logic changes required as a result of the change order. The contractor shall include, as part of each change order proposal, a sketch showing all schedule logic revisions, duration changes, and the relationships to other activities in the approved Initial Schedule. This sketch shall be known as the fragnet for the change. Upon acceptance of the fragnet, the contractor will revise the Initial Schedule or current update. The logic change work required by the change order will be considered incidental to the contract work. No separate payment will be made.

Schedule Revisions to Utility Work. The contractor shall provide the utilities ten days notice when revisions in the schedule of work affect operations of a utility unless previous arrangements have been made with the utility company involved.

If you do not ask for a good schedule, many times you will not get one. Thus, a schedule specification such as the one provided here is always recommended, particularly on projects where problems such as the discovery of contamination are likely.

Suspensions of Work

One of the consequences of the discovery of contamination is that project work may have to be stopped. This may happen because it is no longer safe, because access cannot be provided until remediation is completed, or because the project design must be revised. In any of these circumstances, it may be desirable to provide a method for suspending the contractor's work until the problem can be corrected. The following is the suspension clause used by the federal government. Similar ones are used widely throughout the construction industry.

52.212.-12 Suspensions of Work

(a) The Contracting Officer may order the contractor, in writing, to suspend, delay, or interrupt all or any part of the work of this contract for the period of time that the Contracting Officer determines appropriate for the convenience of the Government.

(b) If the performance of all or any part of the work is, for an unreasonable period of time, suspended, delayed, or interrupted (1) by an act of the Contracting Officer in the administration of this contract, or (2) by the Contracting Officer's failure to act within the time specified in this contract (or within a reasonable time if not specified), an adjustment shall be made for an increase in the cost of performance of this contract (excluding profit) necessarily caused by the unreasonable suspension, delay, or interruption and the contract modified in writing accordingly. However, no adjustment shall be made under this clause for any suspension, delay, or interruption to the extent that performance would have been so suspended, delayed, or interrupted by any other cause, including the fault or negligence of the contractor, or for which an equitable adjustment is provided for or excluded under any other term or condition of this contract.

(c) A claim under this clause shall not be allowed (1) for any costs incurred more than 20 days before the contractor shall have notified the Contracting Officer in writing of the act or failure to act involved (but this requirement shall not apply as to a claim resulting from a suspension order), and (2) unless the claim, in an amount stated, is asserted in writing as soon as practical after the termination of a suspension, delay, or interruption, but no later than the date of final payment under the contract.

Note that this clause addresses the reimbursement a contractor might be entitled to in the event the owner must suspend the project. This clause provides for compensation attributable to a project delay that arises from the suspension. Consequently, this clause would not be compatible with a no-damage-for-delay clause in a contract. This clause and a no-damage-for-delay clause would be in conflict, because one clause would allow reimbursement for delays and the other would not. Note also that while this clause would compensate a contractor for costs incurred as a result of the suspension, it does not allow the contractor profit on these costs.

Liquidated Damages and Time Extension Clauses

One of the consequences of a contractor's failure to complete the contract work on time may be added costs incurred by the owner as a result of the late completion. One way for the owner to recover these added costs is to total the actual expenditures that resulted from the contractor's late completion and subtract this amount from the amounts due the contractor, or otherwise to seek reimbursement from the contractor. Calculating the amount due requires considerable effort in the form of auditing the available financial records, tabulating the results of the audit, and assembling the necessary backup information. If the owner does not have the staff or expertise to conduct such an audit or prepare the necessary documentation, it may have to contract this work out. Also, it is often extremely difficult, if not impossible, to define and prove all costs caused by a delay, such as loss of revenues, user costs, etc. One way to avoid these problems is to estimate what these costs might be before any work is performed and then liquidate the estimated amount into the contract. In other words, rather than calculating the cost of the delay to the owner after the delay is incurred, the owner can estimate these costs before the work is performed and stipulate the amount that would be due in the contract. This stipulated amount is known as *liquidated damages*. The use of a liquidated damages provision is common in construction contracting. Its popularity stems not just from the fact that it makes calculating the owner's delay damages easier, but also because there is the belief that the liquidated damages amount serves as an incentive to the contractor to finish the project on time.

The owner should be careful not to overemphasize the perceived incentive nature of liquidated damages. If the owner, in an effort to increase the contractor's incentive to finish early, increases the liquidated damages amount to the point that it no longer represents a reasonable estimate of the owner's damages in the event of a delay, the liquidated damages become a penalty. An owner cannot penalize or fine a contractor for completing a project late. The owner may recover an amount representing its reasonable estimate of the cost of the project finishing late, but no more. The power to assess penalties is a power that the courts have reserved. It is not a power they have delegated to the owner of a construction project.

The owner has considerable flexibility in the assessment of liquidated damages. For example, if intermediate milestones have been established for the project, liquidated damages may be assessed for these milestones based on a reasonable estimate of the owner's costs if these milestones are not met. If the owner incurs costs because a project is completed late, the owner may also realize some savings when a project is completed early. The owner may establish this amount or some portion of this amount as a bonus or incentive to the contractor for early completion. When combined with a liquidated damages clause, the combined contract provision has sometimes been called a *bonus/penalty* or *incentive/disincentive clause*. While the terms *bonus* and *incentive* are acceptable, beware of the terms *penalty* and *disincentive*. The owner may not assess a penalty, even if a bonus is offered, and disincentives have been looked upon as penalties by some courts. Regardless of the terminology used, any amount set forth in a contract to be assessed the contractor for late completion must have as its basis a reasonable estimate of the owner's damages. It is advisable for the owner to retain the estimate prepared to determine the liquidated damages amount as verification that the liquidated damage truly is a reasonable estimate of the owner's delay costs. The following is an example of an incentive/disincentive clause.

> The following incentive/disincentive will apply to the project completion date. The incentive payment will be limited to a maximum of $100,000.
>
> A. *Disincentive—Failure to Complete in Time.* If the contractor fails to complete the project on or before the project completion date or on or before the authorized extension thereof, the contractor will be charged, for each calendar day that the work shall remain incomplete, the sum of $3,000 per calendar day.
>
> B. *Incentive—Early Completion.* If the contractor completes the project before the contract completion date specified or before the authorized extension thereof, the contractor will be reimbursed, for each calendar day from the time the project is actually completed to the contract time specified or the authorized extension thereof, the sum of $3,000 per calendar day.
>
> Liquidated damages as per the special provision in the proposal for the Schedule of Liquidated Damages will be charged in addition to any disincentive charged for failure to complete in time.

If the owner is serious about completing the project on time, and has included a liquidated damages provision in the contract, the owner must also address situations when the contract time must be extended. This is particularly true when there is a possibility that contamination may be discovered. One of the consequences of discovering contamination on the project site, particularly when the contamination is extensive and/or unexpected, is that the project completion date may be delayed.

Time extensions are generally provided by a contract clause known as a *time extensions clause*. The following is an example of a time extension clause.

> *108.06 Determination of Extension of Contract Time.* The number of working days, calendar days, or the specified completion date of work will be provided in the proposal and will be known as the contract time.
>
> When the contract is on a calendar day basis, the time charge shall begin on the earliest of the following:
>
> 1. The day any work is performed on a pay item, or
> 2. The date specified in the "Notice to Commence Work" under Subsection 108.02.

When the contract completion time is on a specified completion date, it shall be the date on which all work on the project shall be completed.

If the contractor experiences delays to the completion of the project which are unforeseeable and without the fault or negligence of the contractor, the delay will be deemed excusable and the time allowed for performance of the work may be extended according to the following:

1. For delays caused by acts of God, acts of the public enemy, fires, floods, epidemics, quarantine restrictions, strikes, freight embargoes, and unusually severe weather, the contractor may be granted an extension of time, but no additional compensation will be paid.
2. For delays caused by the Department, the contractor may be granted an extension of time according to this Subsection and additional compensation according to Subsection 109.11, entitled "Compensation for Project Delays."

To request an extension of time, the contractor must notify the Engineer according to the following procedures:

1. *Notification of Delay.* Within five calendar days of the occurrence of a delay to the prosecution of the work, the contractor shall notify the Resident Engineer in writing of such a delay and indicate that a request for delay consideration will be filed with the Department.
2. *Procedures Following Notification of Delay.* After notifying the Resident Engineer of the request for delay consideration, the contractor shall keep daily records of all nonsalaried labor, material costs, and equipment expenses for all operations that are affected by the delay.
3. *Procedures for Documenting the Delay.* If the contractor seeks a time extension for delays to any milestone or the contract completion date, the contractor shall furnish the following documentation to the Engineer to enable the Engineer to determine whether a time extension is appropriate under the terms of the contract. Only delays to activities which affect the critical path to milestone dates or the contract completion date will be considered for a time extension.

The contractor shall maintain a daily record of each operation affected by the delay and the station location of the operations affected. Daily records of the operations will also be maintained by the Department. Copies of the contractor's and the Department's daily records will be provided to the respective parties by 9:00 A.M. of the day following the date of record. Each Monday, compare the previous week's daily records with the records kept by the Department. The contractor shall also prepare and submit written reports to the Resident Engineer containing the following information each Monday:

1. Number of days behind schedule.
2. A summary of all operations that have been delayed, or will be delayed.
3. In the case of a compensable delay, the contractor shall explain how the Department's act or omission delayed each operation and the project.
4. Itemize all extra costs being incurred, including:
    (a) How the extra costs relate to the delay and how they are being calculated and measured;
    (b) The identification of all nonsalaried project employees for whom costs are being compiled; and

(c) A summary of time charges for equipment, identified by the manufacturer's number for which costs are being compiled.

Provide written notice to the Engineer within ten calendar days of the results of the comparison of the detailed reports performed each Monday and define any disagreements between specific records.

Failure to meet to review the Department's records or to report disagreements between the records will be considered conclusive evidence that the Department's records are accurate. Recovery of delay costs allegedly incurred before notifying the Resident Engineer that operations have been delayed will not be allowed.

1. Procedures Following Completion of Work Allegedly Delayed—Within 15 calendar days of project completion or phase of work allegedly delayed, whichever is earlier, the contractor shall submit a report to the Resident Engineer containing the following information:

   (a) A description of the operations that were delayed and the documentation and explanation of the reason for the delay, including all reports prepared for the contractor by consultants, if used; and
   (b) An as-built chart showing when work operations were actually performed; and
   (c) A graphic depiction of how the operations were delayed; and
   (d) An item by item measurement and explanation of extra costs requested for reimbursement due to the delay.

All costs shown in the report submitted to the Department must be certified by an accountant.

The Engineer will review the contractor's submission and any reports prepared by the Resident Engineer. A written decision will be provided to the contractor within 60 calendar days of the receipt of the contractor's submission.

In the case of compensable delays, if the Engineer determines that the Department is responsible for delays to the contractor's operations, the Engineer's written decision will reflect the nature and extent of any resulting equitable adjustment to the Contract as contained in Subsection 109.11.

This contract clause accomplishes a number of tasks. First, it defines the basis for measuring contract time and the point at which time charges would begin. Second, it describes the condition under which a time extension would be considered. This clause defines when a time extension would be compensable. In other words, it defines when a contractor might be able to recover additional compensation for a delay. Finally, the clause defines the information required for the contractor to establish its entitlement to a time extension and the timing of required submittals and responses.

A contractor will generally be entitled to a time extension when the project delay that is incurred is neither the contractor's fault nor responsibility, and could not reasonably have been anticipated by the contractor. In many cases, delays that result from the discovery of contamination result in a time extension. This is not always the case, however, particularly when the discovery was not a surprise or responsibility for the discovery was placed on the contractor. In such cases, no time extension may be due. It is the contract, however, that defines excusability and the contractor's entitlement to a time extension.

Termination

The discovery of contamination can jeopardize the very basis for proceeding with a project. This may be because the cost of removing or otherwise handling the contamination is so great that the project is no longer feasible, or because the nature or extent of the contamination is such that the redesign of the project or the remediation will take so long that it does not make sense to continue construction of the project until the problem can be resolved. Whatever the reason, the discovery of contamination may put the owner in a position where it is desirable to terminate the contract. The contract is being terminated not because the contractor's performance has been deficient, but because it no longer makes practical or economic sense for the owner to proceed. In situations such as these, a *termination for convenience clause* is essential. The following is an example of a termination for convenience clause used by the federal government.

> 52.249-2 TERMINATION FOR CONVENIENCE OF THE GOVERNMENT (FIXED PRICE)
>
> (a) The Government may terminate performance of work under this contract in whole or, from time to time, in part if the Contracting Officer determines that a termination is in the Government's interest. The Contracting Officer shall terminate by delivering to the contractor a Notice of Termination specifying the extent of termination and the effective date.
>
> (b) After receipt of a Notice of Termination, and except as directed by the Contracting Officer, the contractor shall immediately proceed with the following obligations, regardless of any delay in determining or adjusting any amounts due under this clause:
>
> (1) Stop work as specified in the notice.
> (2) Place no further subcontracts or orders (referred to as subcontracts in this clause) for materials, services, or facilities, except as necessary to complete the continued portion of the contract.
> (3) Terminate all subcontracts to the extent they relate to the work terminated.
> (4) Assign to the Government, as directed by the Contracting Officer, all rights, title, and interest of the contractor under the subcontracts terminated, in which case the Government shall have the right to settle or to pay any termination settlement proposal arising out of those terminations.
> (5) With approval or ratification to the extent required by the Contracting Officer, settle all outstanding liabilities and termination settlement proposals arising from the termination of subcontracts; the approval or ratification will be final for purposes of this clause.
> (6) As directed by the Contracting Officer, transfer title and deliver to the Government (i) the fabricated or unfabricated parts, work in process, completed work, supplies, and other material produced or acquired for the work terminated, and (ii) the completed or partially complete plans, drawings, information, and other property that, if the contract had been completed, would be required to be furnished to the Government.
> (7) Complete performance of the work not terminated.

(8) Take any action that may be necessary, or that the Contracting Officer may direct, for the protection and preservation of the property related to this contract that is in the possession of the contractor and in which the Government has or may acquire an interest.

(9) Use its best effort to sell, as directed or authorized by the Contracting Officer, any property of the types referred to in subparagraph (6) above; provided, however, that the contractor (i) is not required to extend credit to any purchaser and (ii) may acquire the property under the conditions prescribed by, and a process approved by, the Contracting Officer. The proceeds of any transfer or disposition will be applied to reduce any payments to be made by the Government under this contract, credited to the price or cost of the work, or paid in any other manner directed by the Contracting Officer.

(c) After expiration of the plant clearance period as defined in Subpart 45.6 of the Federal Acquisition Regulation, the contractor may submit to the Contracting Officer a list, certified as to quantity and quality, of termination inventory not previously disposed of, excluding items authorized for disposition by the Contracting Officer. The Contractor may request the Government to remove those items or enter into an Agreement for their storage. Within 15 days, the Government will accept title of those items and remove them or enter into a storage Agreement. The Contracting Officer may verify the list upon removal of the items, or if stored, within 45 days from submission of the list, and shall correct the list, as necessary, before final settlement.

(d) After termination, the contractor shall submit a final termination settlement proposal to the Contracting Officer in the form and with the certification prescribed by the Contracting Officer. The contractor shall submit the proposal promptly, but no later than one year from the effective date of termination, unless extended in writing by the Contracting Officer upon written request of the contractor within this one year period. However, if the Contracting Officer determines that the facts justify it, a termination settlement proposal may be received and acted on after one year or any extension. If the contractor fails to submit the proposal within the time allowed, the Contracting Officer may determine, on the basis of information available, the amount, if any, due the contractor because of the termination and shall pay the amount determined.

(e) Subject to paragraph (d) above, the contractor and the Contracting Officer may agree upon the whole or any part of the amount to be paid because of the termination. The amount may include a reasonable allowance for profit on work done. However, the agreed amount, whether under this paragraph (e) or paragraph (f) below, exclusive of costs shown in subparagraph (f)(2) below, may not exceed the total contract price as reduced by (1) the amount of payments previously made and (2) the contract price of work not terminated. The contract shall be amended, and the contractor paid the agreed amount. Paragraph (f) below shall not limit, restrict, or affect the amount that may be agreed upon to be paid under this paragraph.

(f) If the contractor and the Contracting Officer fail to agree on the whole amount to be paid the contractor because of the termination of work, the Contracting Officer shall pay the contractor the amounts determined as follows, but without duplication of any amounts agreed upon under paragraph (e) above;

(1) For contract work performed before the effective date of termination, the total (without duplication of any items) of—

(i) The cost of this work;
(ii) The cost of settling and paying termination settlement proposals under terminated subcontracts that are properly chargeable to the termination portion of the contract if not included in subdivision (i) above; and
(iii) A sum, as profit on (i) above, determined by the Contracting Officer under 49.202 of the Federal Acquisition Regulation, in effect on the date of this contract, to be fair and reasonable; however, if it appears that the contractor would have sustained a loss on the entire contract had it been completed, the Contracting Officer shall allow no profit under this subdivision (iii) and shall reduce the settlement to reflect the indicated rate of loss.

(2) The reasonable costs of settlement of the work terminated, including—

(i) Accounting, legal, clerical, and other expenses reasonably necessary for the preparation of termination settlement proposals and support data;
(ii) The termination and settlement of subcontracts (excluding the amounts of such settlements); and
(iii) Storage, transportation, and other costs incurred, reasonably necessary for the preservation, protection, or disposition of the termination inventory.

(g) Except for normal spoilage, and except to the extent that the Government expressly assumed the risk of loss, the Contracting Officer shall exclude from the amounts payable to the contractor under paragraph (f) above, the fair value, as determined by the Contracting Officer, of property that is destroyed, lost, stolen, or damaged so as to become undeliverable to the Government or to the buyer.

(h) The cost principles and procedures of Part 31 of the Federal Acquisition Regulation, in effect on the date of this contract, shall govern all costs claimed, agreed to, or determined under this clause.

(i) The contractor shall have the right of appeal, under the Disputes clause, from any determination made by the Contracting Officer under subparagraph (d), (f), or (k), except that if the contractor failed to submit the termination settlement proposal within the time provided in paragraph (d) or (k), and failed to request a time extension, there is no right of appeal. If the Contracting Officer has made a determination of the amount due under paragraph (d), (f), or (k), the Government shall pay the Contracting Officer if there is no right of appeal or if no timely appeal has been taken, or the amount finally determined on an appeal.

(j) In arriving at the amount due the contractor under this clause, there shall be deducted—

(1) All liquidated advance or other payments to the contractor under the terminated portion of this contract;
(2) Any claim which the Government has against the contractor under this contract; and
(3) The agreed price for, or the proceed of sale of materials, supplies, or other things acquired by the contractor or sold under the provisions of this clause and not recovered by or credited to the Government.

(k) If the termination is partial, the contractor may file a proposal with the

Contracting Officer for an equitable adjustment of the price(s) of the continued portion of the contract. The Contracting Officer shall make any equitable adjustment agreed upon. Any proposal by the contractor for an equitable adjustment under this clause shall be requested within 90 days from the effective date of termination unless extended in writing by the Contracting Officer.

(1) The Government may, under the terms and conditions it prescribes, make partial payments and payments against costs incurred by the contractor for the terminated portion of the contract, if the Contracting Officer believes the total of these payments will not exceed the amount to which the contractor will be entitled.

(2) If the total payment exceeds the amount finally determined to be due, the contractor shall repay the excess to the Government upon demand, together with interest computed at the rate established by the Secretary of the Treasury under 50 U.S.C. App. 1215(b)(2). Interest shall be computed for the period from the date the excess payment is received by the contractor to the date the excess is repaid. Interest shall not be charged on any excess payment due to a reduction in the contractor's termination settlement proposal because of retention or disposition, or a later date determined by the Contracting Officer because of the circumstances.

(l) Unless otherwise provided in this contract or by statue, the contractor shall maintain all records and documents relating to the terminated portion of this contract for three years after final settlement. This includes all books and other evidence bearing on the contractor's costs and expenses under this contract. The contractor shall make these records and documents available to the Government, at the contractor's office, at all times, without any direct charge. If approved by the Contracting Officer, photographs, microphotographs, or other authentic reproductions may be maintained instead of original records and documents.

This clause has three basic parts. The first part includes a definition of when the owner may want to terminate the contract. The second part describes the mechanism by which the termination will be effected. The third and last part describes how the contractor will be paid for the work that was performed before the contract was terminated. This last part is essential to the smooth functioning of the clause. Note that this clause specifically prohibits the recovery of anticipated profits. The contractor will be entitled to a markup for profit on the costs that were incurred, but not recovery of the profit it might have realized had the project been pursued to completion.

In situations where the contract is terminated for failure of the contractor to fulfill the contract requirements, the termination is known as *default termination* or *termination for cause*. Such a termination is governed by a *default termination clause*. The following is an example of such a clause.

108.10 Default of Contract. If the Contractor:

a. Fails to begin the work under the contract within the time specified in the Notice to Proceed or

b. Fails to perform the work with sufficient workmen and equipment or with sufficient materials to assure the prompt completion of said work or
c. Performs the work unsuitably or neglects or refuses to remove materials or to perform anew such work as may be rejected as unacceptable and unsuitable or
d. Discontinues the prosecution of the work or
e. Fails to resume work which has been discontinued within a reasonable time after notice to do so or
f. Becomes insolvent or is declared bankrupt or commits any act of bankruptcy or insolvency or
g. Allows any final judgment to stand against him unsatisfied for a period of ten days or
h. Makes an assignment for the benefit of creditors without authorization by the Department, or
i. For any other cause whatsoever, fails to carry on the work in a manner acceptable to the Department, the Engineer will give notice in writing to the contractor and his surety of such delay, neglect or default.

If the contractor or Surety, within a period of ten days after such notice, does not proceed in accordance therewith, then the Commissioner will, upon written notification of the fact of such delay, neglect or default and the contractor's failure to comply with such notice, have full power and authority without violating the contract, to take prosecution of the work away from the contractor. The Department may use any or all materials and equipment as may be suitable and may enter into an Agreement for the completion of said contract according to the terms and provisions thereof or use such other methods as in the opinion of the Engineer will be required for the completion of the contract in an acceptable manner.

All costs and charges incurred by the Department, together with the cost of completing the work done under the contract, will be deducted from monies due or which may become due the contractor. If such expense exceeds the sum which would have been payable under the contract, the contractor and the Surety shall be liable and shall pay to the Department the amount of such excess.

The following is the default termination clause used by the federal government.

52.249-10 DEFAULT (FIXED-PRICE CONSTRUCTION)

(a) If the contractor refuses or fails to prosecute the work or any separable part, with the diligence that will insure its completion within the time specified in this contract including any extension, or fails to complete the work within this time, the Government may, by written notice to the contractor, terminate the right to proceed with the work (or the separable part of the work) that has been delayed. In this event, the Government may take over the work and complete it by using materials, appliances, and plant on the work site necessary for completing the work. The contractor and its sureties shall be liable for any damage to the Government resulting from the contractor's refusal or failure to complete the work within the specified time, whether or not the contractor's right to proceed with the work is terminated. The liability includes any increased costs incurred by the Government in completing the work.

(b) The contractor's right to proceed shall not be terminated nor the contractor charged with damages under this clause, if—

  (1) The delay in completing the work arises from unforeseen causes beyond the control and without the fault or negligence of the contractor. Examples of such causes include (i) acts of God or of the public enemy, (ii) acts of the Government in either its sovereign or contractual capacity, (iii) acts of another contractor in the performance of a contract with the Government, (iv) fires, (v) floods, (vi) epidemics, (vii) quarantine restrictions, (viii) strikes, (ix) freight embargoes, (x) unusually severe weather, or (xi) delays of subcontractors or suppliers at any tier arising from unforeseeable causes beyond the control and without the fault or negligence of both the contractor and the subcontractor suppliers; and

  (2) The contractor, within 10 days from the beginning of any delay (unless extended by the Contracting Officer), notifies the Contracting Officer in writing of the causes of delay. The Contracting Officer shall ascertain the facts and the extent of delay. If, in the judgement of the Contracting Officer, the findings of fact warrant such actions, the time for completing the work shall be extended. The findings of the Contracting Officer shall be final and conclusive on the parties, but subject to appeal under the Disputes clause.

(c) If, after termination of the contractor's right to proceed, it is determined that the contractor was not in default, or that the delay was excusable, the rights and obligations of the parties will be the same as if the termination had been issued for the convenience of the Government.

(d) The right and remedies of the Government in this clause are in addition to any other rights and remedies provided by law or under this contract.

Though this example focuses more on the issue of time, paragraph (c) is of particular importance. It converts a termination for default into a termination for convenience in the event that the contractor was determined to be not in default. It is very common for a contractor to dispute a termination for default. It is also common for such a dispute to end up in the courts. The reason this issue is so contentious is that the very survival of a contractor as a profitable business may be jeopardized by a termination for default. Sureties are reluctant to bond companies that have been terminated for default, and owners are reluctant to hire these contractors. Consequently, unless the contractor can have the default determination by the owner overturned, the contractor's chances for business success may be markedly reduced. Given the drastic nature of the alternative, a good fight seems to be well worth the effort, particularly if there is some hope of success. And there is some hope. Contractors have successfully overturned default terminations on many occasions. When this happens, the damages that the owner might be exposed to from a wrongful termination can vastly exceed the costs the contractor incurred performing the contract work. These damages may include costs related to the contractor's loss of bonding capacity or loss of business, the harm done to the contractor's reputation, and litigation expenses. If, however, the termination is converted to a termination for convenience, then the damages to which the contractor would be entitled are limited to the costs allowed by the termination for convenience clause. Damages related to lost business or profits would not be allowed.

Dispute Prevention and Resolution

In addition to the partnering provisions discussed earlier in this chapter, there are contractual steps that can be taken to try and reduce the incidence of disputes and to help resolve these disputes quickly, before they escalate into a full-fledged battle.

One technique that has been advocated is the escrow of bid documentation. This is done to provide some comfort to the owner that additional payments being considered for added work were not already included in the contractor's bid. Escrowed bid documents can also be used to establish overhead rates, and to prove that the amounts bid for unit-price items in the bids are balanced. An unbalanced unit price is a unit price that is significantly greater than or less than the reasonable cost of performing a particular unit of work. Owners do not usually like unbalanced unit prices. In some cases, such unbalancing may reflect errors in the plans and specifications and underestimated quantities. If the contractor finds these problems and, rather than notifying the owner, increases its unit price for the underestimated item, it is possible that the price ultimately paid by the owner to perform the work may be higher than the price offered by the next lowest bidder. If the owner detects such a situation, the contractor's bid could be rejected because it is "materially unbalanced." Escrowed bid documents help the contractor establish that the price was not in fact unbalanced or the owner establish that the bid was unbalanced.

The following clause has been used to provide for the escrow of bid documentation.

> **Escrow of Bid Documentation.** If specified, the bidder shall submit with the bid Proposal a legible copy of bid documentation used to prepare the Contract bid. The documentation of the successful bidder will be placed in escrow with a banking institution or other bonded document storage facility and preserved by that institution/facility as specified in the following sections.
>
> A. *Submittal and Return of Bid Documentation.* The bidder shall submit the bid documentation in a sealed container. The container shall be clearly marked "Bid Documentation" and show on the face of the container the bidder's name and address, the date of submittal, the Project Number, and the Contract Number. Within seven calendar days of award of the Contract, the sealed containers of the remaining unsuccessful bidders will be returned to them by the Department.
>
> B. *Affidavit.* In addition to the bid documentation, the bidder shall submit an affidavit, signed under oath by a representative of the bidder authorized to execute bidding Proposals, listing each bid document submitted by author, date, nature, and subject matter. The affidavit shall attest (1) that the affiant has personally examined the bid documentation, (2) that the affidavit lists all of the documents relied upon by the bidder in preparing its bid for the project,

and (3) that all such bid documentation is included in the sealed container submitted to the Department.

C. *Duration and Use.* After execution of the Contract, the Department and the contractor will jointly deliver the sealed container and affidavit to a banking institution or other bonded document storage facility selected by the Department for placement in a safety deposit box, vault, or other secure accommodation.

The document depository agreement shall reflect that the bid documentation and affidavit shall remain in escrow during the life of the Contract or until the Department is notified of the contractor's intention to file a claim or initiate litigation against the Department related to the Contract. Upon mutual agreement or with notification of the contractor's intention to file a claim, or initiation of litigation against the Department, the Department may obtain the release and custody of the bid documentation. In the absence of such action and provided that the contractor signs the final Standard Release Form, the Department will instruct the document depository to release the sealed container to the contractor.

In accordance with the Contractor's representation that the sealed container placed in escrow contains all of the materials relied upon in preparing its bid, the contractor agrees to waive the right to use any bid documentation other than that placed in escrow in disputes arising out of the contract.

D. *Refusal or Failure to Provide Bid Documentation.* Failure to provide documentation shall render the bid non-responsive.

E. *Confidentiality of Bid Documentation.* The bid documentation and affidavit in escrow are, and shall remain, the property of the contractor. The Department has no interest in, or right to, the bid documentation unless notification of the intention to file claim is received or litigation ensues between the Department and contractor. In the event of such notification or litigation, the bid documentation and affidavit will become the property of the Department until complete resolution of the claim or litigation is achieved. These materials, and all copies made by the Department, will be returned to the contractor at the conclusion of litigation, or final resolution of all outstanding claims, upon execution of a final release. The Department will make every reasonable effort to ensure that the bid documentation remains confidential within the Department and will not be made available to anyone outside the Department, or used by a former Department employee.

F. *Cost and Escrow Instructions.* The cost of the escrow documentation depository will be borne by the Department. The Department will provide escrow instructions to the document depository consistent with this clause.

G. *Payment.* There will be no separate payment for compilation of the data, container or cost of verification of the bid documentation. All costs shall be included in the overall Contract bid price.

Another technique that has been used to encourage disputes to be resolved during the performance of the project is the use of disputes review boards. These boards are discussed in greater detail in Chap. 9, but the following clause or something similar might be inserted in the contract to effect their use.

Disputes Resolution

In order to assist in the resolution of disputes or claims arising out of the work of this contract and all related contracts for the project, the Department has provided for the establishment of a Disputes Review Board, hereafter called the "Board." The Board has been added to the disputes resolution process to be brought into play when the normal Department-Contractor dispute resolution process is unsuccessful and prior to a formal adoption of position or filing of claim and instigation of litigation by either party.

a. *Disputes.* Disputes, as used in this Subsection, will include disagreements, claims, counterclaims, matters in question, and differences of opinion between the Department and the contractor:

1. On matters related to the work and to change or extra work orders, including:
    (a) Interpretation of the Contract
    (b) Costs
    (c) Time for performance
2. And on other subjects requested by either party.

b. *Resolution Procedure.* The following procedure shall be used for disputes resolution:

1. If seven calendar days have expired since the contractor provided the signed written notice of protest and the contractor continues to object to the decision or order of the engineer, the contractor shall request, in writing, written instructions from the engineer. The contractor's request shall include the following:
    (a) The date of the protested order,
    (b) The nature and circumstances which caused the protest,
    (c) The contract provisions that support the protest,
    (d) The estimated dollar cost, if any, and if available, of the protested work and how the estimate was determined, and
    (e) An analysis of the progress schedule showing the schedule change or disruption if the contractor is asserting a schedule change or disruption.
2. The engineer shall respond, in writing, to the contractor's written request within 15 calendar days.
3. Within 30 calendar days after receipt of the engineer's written instructions, the contractor shall, if the contractor still objects to such instruction, appeal to the Board.
4. The contractor and the Department shall each be afforded an opportunity to be heard by the Board and to offer evidence. Either party furnishing any written evidence or documentation to the Board must furnish copies of such information to the other party a minimum of 15 calendar days prior to the date the Board sets to convene the hearing for the dispute. Either party shall produce such additional evidence as the Board may deem necessary to an understanding and determination of the dispute and furnish copies to the other party.
5. The Board's recommendation toward resolution of a dispute will be given in writing to both the Department and the contractor. The recommenda-

tions will be based on the contract provisions and the actual costs and or the time incurred.
   6. Within 30 calendar days of receiving the Board's recommendations, both the Department and the contractor shall respond to the other in writing signifying that the dispute is either resolved or remains unresolved.
   7. Although both parties should place weight upon the Board's recommendations, the recommendations are not binding. Either party may appeal a recommendation of the Board to them for reconsideration. However, reconsiderations shall only be allowed when there is new evidence to present.
   8. If the Department and the contractor are able to resolve their dispute with the aid of the Board's recommendations, the Department will promptly process any contract changes.
   9. In the event the Board's recommendations do not resolve the dispute, all records, and written recommendations, including any minority reports, will be admissible as evidence in any subsequent litigation.
c. *Litigation*
   1. Submittal of dispute to the Board shall be a condition precedent to filing for litigation in a court of law unless the Department and the contractor have agreed to directly proceed to Subsection 105.222, Claims.
   2. Claims, counterclaims, disputes, and other matters in question between the Department and the contractor that are not resolved will be decided in court, which shall have exclusive jurisdiction and venue over all matters in question between the Department and the contractor.
   3. The Contract shall be interpreted and constructed in accordance with the laws of the State of Maine.
d. *Purpose and Function of the Board.* The Board will be an advisory body created to assist in the resolution of claims, disputes or controversy between the contractor and the Department in order to prevent construction delay and possible court litigation.

   The Board will consider disputes referred to it, and furnish recommendations to the Department and the contractor to assist in the resolution of the differences between them. The Board will essentially be making nonbinding findings and recommendations and provide special expertise to assist and facilitate the resolution of disputes.
e. *Board Members.* A Disputes Review Board is in place for the contract. This Board was formed from candidates selected from a list of candidates created jointly by the Department, Maine's Associated General Contractors and the contractor.

   At the contractor's option, the present Board may be appointed to this Contract or a new member, selected by the contractor from a list on file at the Department will be appointed to replace the present contractor's member. The contractor's member and the Department's member will select a third member from the list to act as chairman if they so desire. All three Board members shall be acceptable to the contractor and the Department.

   In case a member of the Board needs to be replaced, the replacement member will be appointed in the same manner as the replaced member was appointed. The appointment of a replacement Board member will begin promptly upon determination of the need for replacement and shall be completed within 30 calendar days.

Service of a Board member may be terminated at any time with not less than 30 calendar days notice as follows:

1. The Department may terminate service of the Department appointed member.
2. The contractor may terminate service of the contractor appointed member.
3. The third member's services may be terminated only by the Agreement of the other two members.
4. By resignation of the member.
5. Termination of a member will be followed by appointment of a substitute as specified above.

No member shall have a financial interest in the contract, except for payment for services on the Board. No member shall have been employed by either party within a period of two years prior to award of this contract, except that, services as a member of other Dispute Review Boards on other contracts will not preclude a member from serving on the Board of this contract.

The Board members will be especially knowledgeable in the field of construction of the type covered by the Contract and shall discharge their duties based on the facts and conditions related to the matters under consideration and the provisions of the Contract.

f. *Board Operation.* The Board will formulate its own rules of operation to keep abreast of the construction development, the members shall meet quarterly with the members of the Board and with representatives of the Department and the contractor at the Project site.

In order for this Disputes Review process to be successful, Board members must act as neutrals and shall conduct their discussion accordingly. Ex parte communications between Board members and representatives of the parties is expressly prohibited.

When either the Department or the contractor have an issue which cannot wait for the next scheduled quarterly Board meeting, they must mutually agree to request the Board to meet.

g. *Service Agreements and Compensation.* Service agreements with Board members shall be negotiated by the Department.

All the Service agreements shall be executed in forms acceptable to the Department.

The Department will compensate the Board members and pay their expenses for quarterly meetings.

Compensation for the Board members and expenses, and the expenses of operation of the Board, for any additional meetings called to resolve a dispute shall be shared equally by the Department and contractor.

The Department, through the Engineer, will provide administrative services, such as conference facilities and secretarial services, to the Board and the Department will bear the costs for this service.

The Engineer will evaluate all protests provided the procedures in this Section are followed. If the Engineer determines that a protest is valid, the Engineer will adjust payment for work according to Subsection 109.03 or 109.04 or time according to Subsection 109.11.

By failing to follow the procedures of this Subsection and Subsection 105.22, the contractor completely waives any claim for protested work.

The contract clauses provided in this book are presented as examples. They may or may not apply to your particular project. It is always a good idea to have your contract documents reviewed prior to use. This review could be made a part of the design or constructability reviews or a separate review. Reviewers should not only be experienced with drafting contract language for construction projects, but, in situations where the contract documents must deal with the potential for discovering contamination, should also have experience with similar projects and contract formats. Finally, the contract documents should be reviewed by qualified legal counsel who has extensive experience in construction law.

Chapter 5

Immediate Actions upon Discovery of Contamination

While excavating for the footing of a bridge pier, a backhoe operator uncovered the capped end of a large-diameter pipe. In the process of excavating the area, the backhoe operator inadvertently dislodged the cap at the end of the pipe. Upon examination, the operator noticed a yellow, foul-smelling sludge oozing from the end of the pipe. The operator immediately halted excavation activities and notified the project superintendent. The project owner was informed, the area was fenced off, and experts were brought in to evaluate the yellow sludge. The results of testing showed that the sludge was contaminated with chromium wastes. Ultimately, access to a substantial portion of the project site was restricted while a separate contractor was brought in to remove the abandoned and long-forgotten drainage system of a chrome plating plant that formerly occupied the project site.

While it might have been preferable for the operator to have been more careful and avoided jarring loose the cap in the first place, the actions taken upon the discovery of potential contaminants were correct. Work that might have disturbed conditions further was halted. The appropriate personnel were informed. Action was taken to identify the potential contaminant. Ultimately, the chromium wastes were removed by a competent contractor experienced with the handling of toxic materials. This example also illustrates the basic steps in response to the discovery of contaminants on a construction site. These steps include discovery, notification, addressing safety and health considerations, containment, evaluation, and documentation.

Discovery

The first action associated with finding contamination on a project is the discovery itself. The discovery may be accidental and unexpected, or it may be anticipated. Whether it was a surprise or anticipated, however, the timing of the discovery is essential to minimizing the health and safety hazards and the cost and time impacts of the discovery. Imagine, for example, what would have happened if the backhoe operator, rather than stopping his or her work, had continued excavating and had mixed the chromium wastes with the other excavated materials, used these mixed materials as fill and spread the contamination to other areas of the site, or, worse yet, trucked the material to other sites for use as fill. If the contaminated materials had gotten onto equipment tracks and tires, they could have been spread throughout the site, and to local roadways. They also could have been washed into nearby streams and lakes. Fortunately, the backhoe operator quickly recognized an unusual and potentially hazardous condition and stopped work to prevent further spreading of the contaminants.

There are a number of ways to enhance the chances that project personnel will respond promptly and correctly to the discovery of potentially hazardous materials. The first and perhaps most successful approach is to assure that project personnel receive adequate training. Training is crucial to identifying the presence of potential contaminants and increasing the chances that the hazard will be recognized before the health and safety of the project personnel and public is threatened and before the contaminants are spread throughout the project or even outside the project boundaries. Training is particularly crucial in situations where the contaminant is not easily detected by the normal senses. While the presence of a yellow sludge or a powerful gasoline smell would immediately strike most people as being unusual and potentially dangerous, the presence of lead in paint being removed as part of a restoration project might not be so obvious a problem. Only by assuring that crews are adequately trained regarding contaminants likely to be encountered can contractors and owners be confident that potential problems will be identified before the contaminants are spread more widely throughout the project. Strategies for determining the contaminants that are likely to be encountered on a site are discussed in Chap. 1.

Training can come from many sources. Workers and their supervisors can be sent to seminars, or instructors can be hired to provide training in house. In some cases, usually when contamination on the site is anticipated or has already been identified, it may make sense to mandate a certain basic level of training, knowledge, and understanding. In such cases, site access may be restricted to those that have received the training. Such restricted site access may also be mandated by law, permit, or contract. Training might also be provided inexpensively by knowledgeable and experienced personnel from the contractor's staff, a subcontractor, the owner, the designer, or an owner's representative. It might be provided at the construction site, a local hotel or training facility, or at facilities provided by one of the project's participants.

Different people learn in different ways. Some learn by reading books, manuals, or notes. Some learn from pictures and graphics. Some learn by hearing. Almost everyone, however, learns by doing. Design training programs to include all these methods of learning. Provide typed notes in outline form, with plenty of white space, so that participants can keep notes during the training. Provide some written material to serve as a reference source at a later date. Have someone present the material, in either lecture or seminar format. A seminar format is better, though it may be inappropriate for very basic material and large groups. Use as many pictures and graphics as possible. Videotapes may also be available on particular topics. Also, consider videotaping any training for later use. Almost nothing, however, is more beneficial than real examples, samples, case studies, and practical exercises and problems. For example, if workers are being instructed in how to recognize friable asbestos, have samples available (adequately contained, of course) for viewing.

In addition to providing recognition training, teach personnel how to respond. In other words, make sure they know to whom to report any potential contamination and also how to handle potential contaminants that may be discovered.

A second way to help ensure that workers on the project site will recognize and respond correctly to the discovery of hazardous materials is to hire people who already have some experience. Hiring those with experience may also reduce the amount of training that is required.

Vigilance is also important, particularly when contamination is suspected or expected. It is important for project supervisory and management personnel to get out on the job site. They should not lock themselves in their offices. While this admonition would be true even if discovering contamination was not likely, it can be particularly true on projects where the discovery of contaminants is a possibility. Sometimes there is simply no substitute for an experienced and knowledgeable eye.

Awareness based on research can also help increase the chances of detecting contamination before it is spread throughout the project. For example, simply walking through the project site can alert one to potential hazards such as the presence of asbestos-containing materials, lead-based paint, and other obvious problems. With a little more research effort, the presence of other potential contaminants might also be identified. For example, research into the previous ownership of the chromium-contaminated site described at the beginning of this chapter would have alerted the project participants to the possibility that chromium wastes might be present on the site.

Notification

Once contamination is discovered, it is essential that everyone on the project who needs to know is informed. The contract documents should give some guidance in this respect. Usually, the contractor is obligated by contract to inform the owner in writing whenever a problem is encountered, particularly any problems for which the contractor might ultimately seek added compensation or a time extension. In addition, the contract often requires that this written notice contain particular information such as the time and date of discovery, a description of the problem, and the location. A typical notification letter might read as follows:

Dear Owner (or Owner's Representative):

At 2:00 PM on May 2, 1996, our backhoe operator, Bob Smith, discovered a yellow liquid oozing from the end of a pipe uncovered while excavating for the Pier No. 1 foundation. Mr. Smith stopped work immediately and notified his foreman. I notified you orally of the problem yesterday at approximately 3:30 PM.

The pipe discovered in the excavation from which the yellow ooze is flowing is not shown on the plans and we do not know what the pipe contains. You have directed us to halt our operations in the vicinity of Pier No. 1 until the yellow ooze is evaluated. We have stopped work as directed. Please note that Pier No. 1 excavation work is critical and every day that this work is held up represents a delay to the project completion date. We will seek a time extension for this delay and appropriate additional compensation. While the full cost impact of this discovery will not be known until the actions needed to correct the problem are identified, our field and home office overhead costs are currently running at approximately $2,500 per day.

Our resources are at your disposal. Please let us know how we can help resolve this problem.

Sincerely,
The Contractor's Project Manager

For some contracts even this notice letter would be inadequate, but it addresses the most important points. It states when and where the problem was discovered. It sets forth how the owner was notified of the problem. It describes how the contractor is addressing the problem. It also describes how the problem is affecting the project and what the time and cost consequences might be. This last aspect of a notification letter is often omitted, but the information is very helpful to the owner or the owner's representative. It is important for two reasons: It helps the owner understand the urgency of the problem and gives the problem the appropriate priority; and it helps the owner convey to others the importance of quick resolution. For example, knowing the likely cost impact of the discovery of contamination will help the owner determine or justify the payment of an overtime premium to the testing and design firms that must evaluate the potential contamination and figure out how to deal with the problem. Many times it is not the contamination itself that causes the bulk of the cost and time effects, but the time taken to evaluate the problem and develop a fix. Giving the owner some idea of the cost and time consequences of the discovery will help the owner in deciding on the most economical fix.

Also, it should be noted that a contractual requirement to provide written notice to the owner does not preclude oral notice. In fact, in a situation where potential contamination is discovered, it is important to give notice to all involved parties as quickly as possible. This will not only get everyone working on a fix, it will allow the other parties to give direction regarding immediate actions.

Safety and Health Response

In the event of the discovery of contamination, one of the most important concerns is the safety and health of all workers or other individuals who may have been subject to exposure. In the example at the beginning of the chapter, the equipment operator may have been exposed to the contaminants that were discovered. As soon as responsible parties recognize that this situation might involve hazardous materials, the equipment operator and all other individuals who might have been exposed should undergo a health check to ensure that no one has been harmed or is in danger from the hazardous material. Once the contaminant has been identified, appropriate follow-up health checks should be performed on anyone who might have been exposed.

Projects with even the remotest possibility of hazardous material should have in place a *standard operating procedure* (SOP) for safety and health response and actions in the event of the discovery of potentially hazardous material. This SOP should specify the appropriate actions to be taken, including health checks and area security from the human safety standpoint. It should be remembered that securing the area of potential contamination may not be a complete response. For instance, in the example of the backhoe operator, the equipment should be quarantined until it is determined what the exact situation is. Following that, the equipment may have to be decontaminated. Therefore, when establishing an SOP for health and safety, give thoughtful consideration to all possible ramifications of the discovery of hazardous material.

Merely establishing an SOP does not mean that personnel will be aware of it or understand what it contains or how it should be implemented. The SOP should be included during the training of project personnel. Furthermore, the safety and health issues should be discussed at the contractor's safety meetings with job-site personnel. Normally, these meetings are included as part of the routine job "tool box" meetings. In addition to safety meetings, all significant SOP actions should be prominently posted at the work site, so that they do not have to be "searched for" in the frenzy that may follow the discovery of hazardous materials.

The presence or discovery of hazardous materials on a project site also triggers certain regulatory requirements. The federal government agencies that have responsibility in this area are OSHA and the EPA. OSHA's regulations regarding the employer's responsibilities when hazardous materials are present on the site are found in the *Hazard Communications Standard* (HCS). The EPA is more concerned with the generation and disposal of hazardous wastes as addressed by the Resource Conservation and Recovery Act and attendant regulations. In addition to these basic regulations and acts, detailed regulations have also been developed for particular materials or problems, such as asbestos and underground storage tanks.

OSHA's *Hazardous Communications Standard* focuses primarily on hazardous chemicals in the workplace. Its primary intent is to ensure that the hazards of all chemicals produced, imported into, or used in American workplaces are evaluated, and that this information is disseminated to employers and the employees who will be exposed to these chemicals. The HCS is provided in App. C.

Responsibility for meeting the requirements of the HCS falls to two groups. The first group includes chemical manufacturers and importers. The second group is employers. On a construction project, employers include the owner with representatives on the site, the contractors and subcontractors with workforces performing work on the site, and others providing technical, consulting, inspection, testing, or other services to the site.

The chemical manufacturers and importers are responsible for identifying the hazards associated with particular chemicals. Each chemical must be evaluated for its potential to cause adverse health effects and its potential to pose physical hazards such as flammability. Several sources list hazardous chemicals, including the following:

1. 29 CFR 1910, Subpart Z, Toxic and Hazardous Substances, OSHA
2. Threshold Limit Values for Chemical Substance and Physical Agents in the Work Environment, America Conference of Governmental Industrial Hygienists (ACGIH)
3. National Toxicology Program (NTP), Annual Report on Carcinogens
4. International Agency for Research on Cancer (IARC), Monographs

The chemical producers and importers are then required by the HCS to communicate the results of their evaluation of the hazards represented by particular chemicals to their own employees and to the employers who use these chemicals. This information is conveyed to others via a document known as a *Material Safety Data Sheet* (MSDS). An example of the basic form of such a sheet is shown in Fig. 5.1. The MSDS contains the basic information needed both to understand what the hazard is and the actions that are needed to minimize or prevent exposure. The MSDS form is required to be provided with each purchase or shipment of a hazardous chemical by the producer or importer of the chemical.

In addition to such basic information as the name, address, and telephone number of the manufacturer or the importer, the MSDS lists the hazardous chemical itself and the concentrations or exposure levels at which the chemical becomes a hazard. The MSDS also lists the basic physical properties of the chemical, including boiling and melting points and recognizable traits such as appearance and odor. With regard to specific health hazards, the MSDS describes how the chemical can be brought into the body (inhalation, ingestion, etc.), the short- (acute) and long-term (chronic) health hazards, and whether the chemical is a known carcinogen (cancer-causing agent).

The health hazard section of MSDS identifies the likely symptoms associated with exposure, and emergency and first-aid procedures. The form also lists any medical conditions that might be worsened by exposure to the chemical.

Material Safety Data Sheet May be used to comply with OSHA's Hazard Communication Standard, 29 CFR 1910.1200. Standard must be consulted for specific requirements.	**U.S. Department of Labor** Occupational Safety and Health Administration (Non-Mandatory Form) Form Approved OMB No. 1218-0072	
IDENTITY *(As Used on Label and List)*	*Note: Blank spaces are not permitted. If any item is not applicable, or no information is available, the space must be marked to indicate that.*	

Section I

Manufacturer's Name	Emergency Telephone Number
Address *(Number, Street, City, State, and ZIP Code)*	Telephone Number for Information
	Date Prepared
	Signature of Preparer *(optional)*

Section II — Hazardous Ingredients/Identity Information

Hazardous Components (Specific Chemical Identity; Common Name(s))	OSHA PEL	ACGIH TLV	Other Limits Recommended	% *(optional)*

Section III — Physical/Chemical Characteristics

Boiling Point		Specific Gravity (H_2O = 1)	
Vapor Pressure (mm Hg.)		Melting Point	
Vapor Density (AIR = 1)		Evaporation Rate (Butyl Acetate = 1)	
Solubility in Water			
Appearance and Odor			

Section IV — Fire and Explosion Hazard Data

Flash Point (Method Used)	Flammable Limits	LEL	UEL
Extinguishing Media			
Special Fire Fighting Procedures			
Unusual Fire and Explosion Hazards			

(Reproduce locally)   OSHA 174, Sept. 1985

Figure 5.1 Material Safety Data Sheet.

Section V — Reactivity Data

Stability	Unstable		Conditions to Avoid	
	Stable			

Incompatibility *(Materials to Avoid)*

Hazardous Decomposition or Byproducts

Hazardous Polymerization	May Occur		Conditions to Avoid	
	Will Not Occur			

Section VI — Health Hazard Data

Route(s) of Entry: Inhalation?   Skin?   Ingestion?

Health Hazards *(Acute and Chronic)*

Carcinogenicity: NTP?   IARC Monographs?   OSHA Regulated?

Signs and Symptoms of Exposure

Medical Conditions
Generally Aggravated by Exposure

Emergency and First Aid Procedures

Section VII — Precautions for Safe Handling and Use

Steps to Be Taken in Case Material Is Released or Spilled

Waste Disposal Method

Precautions to Be Taken in Handling and Storing

Other Precautions

Section VIII — Control Measures

Respiratory Protection *(Specify Type)*

Ventilation	Local Exhaust		Special	
	Mechanical *(General)*		Other	
Protective Gloves			Eye Protection	

Other Protective Clothing or Equipment

Work/Hygienic Practices

Figure 5.1 *(Continued)*

Two areas of concern with regard to any hazardous chemicals are handling and methods for controlling exposure. The MSDS addresses handling, including how to deal with releases or spills, disposal methods, and storing. The MSDS also indicates whether handling and use require respiratory protection, gloves, eye protection, protective clothing, the use of special equipment, or other handling concerns.

The employer's responsibilities under HCS are many. Given the number of likely employees on a construction site and the administrative burdens associated with HCS compliance, it may be a good idea to coordinate compliance efforts on the site. This might be accomplished in the contract by assigning a single organization this coordination responsibility; by addressing this concern in the partnering session; or simply by getting together and working out an informal joint approach. Such a coordination effort has the added benefit of keeping other parties on the site informed about the hazardous chemicals present on the site.

Employers are first required to identify and list hazardous chemicals in their workplaces. They are then required to obtain MSDSs and appropriate labels for each hazardous chemical identified. Employers are then required to develop a written *Hazard Communication Program* (HCP). The objective of this program is to make employees aware of the hazardous chemicals on the site. The final requirement for employers with regard to the HCS is to implement or execute their written HCP.

The HCP must address the following topics in writing:

1. Container labeling.
2. Collection and posting of MSDSs.
3. Provide a list of hazardous chemicals on the site.
4. Describe how the employee is to be educated regarding the hazards associated with the chemicals that will be encountered or used in the performance of their work. Since the HCS requires that everyone on site be informed, the HCP should also describe how others on the site will be informed.
5. The HCP must be made available to all employees.

The HCS requires that basic information be provided and training conducted for every employee who will be exposed to the hazardous chemicals. The information and training must cover the following:

1. The hazard communication standard and the requirements of the standard
2. The components of the hazard communication program in the employees' workplaces
3. Operations in work areas where hazardous chemicals are present
4. Where the employer will keep the written hazard evaluation procedures, communications program, lists of hazardous chemicals, and the required MSDS forms

5. How the hazard communication program is implemented in the workplace, how to read and interpret information on labels and the MSDS, and how employees can obtain and use the available hazard information
6. The hazards of the chemicals in the work area
7. Measures employees can take to protect themselves from the hazards
8. Specific procedures put into effect by the employer to provide protection, such as engineering controls, work practices, and the use of personal protective equipment
9. Methods and observations—such as visual appearance or smell—that workers can use to detect the presence of a hazardous chemical to which they may be exposed

A sample Hazardous Communication Program provided by OSHA is included in App. E. In fact, OSHA is an excellent source of information on the topics of the HCS and HCP and hazardous chemicals in general. OSHA has an extensive catalog of audiovisual materials on such subjects as "Health Hazard Recognition in Construction" and "Hazardous Substances." OSHA also offers audiovisual presentations concerning the HCS for the construction industry. A sample written HCP is also available on a computer diskette.

The HCS addresses hazardous chemicals on site. OSHA and the EPA also address the presence of hazardous waste operations taking place on construction projects. The approach to regulation varies to some degree based on the type of hazardous waste that has been encountered. In general, OSHA addresses the risks to employees represented by hazardous waste operations. The EPA is more concerned with the actual procedures used for removal and disposal.

OSHA's general requirements regarding hazardous waste operations are similar to the requirements of the HCS, and are summarized as follows:

1. The hazardous waste site employer—which could be the owner, contractor, or both, depending on the language of the contract agreements between the parties—is required to prepare a safety and health program. This program should address such things as the identification of hazards, evaluation of these hazards, how safety and health concerns will be addressed in light of these hazards, and emergency response procedures.
2. An evaluation of the site to identify potential hazards and possible means of addressing these hazards.
3. Development of a site control program to describe how employees will be protected from hazards. This program would include the following:
    a. A site map
    b. The identification of site work zones
    c. Site communications
    d. Safe work practices
    e. Location of medical assistance
4. Training for all employees who are either involved in the hazardous waste operations or who could be a part of the response to an emergency. Some

basic level of training may be necessary for everyone else on the site, just so they understand postings and emergency procedures.

Training at an "awareness" level should also be provided for "first responders," those employees who are likely to be the first to witness or discover a hazardous substance and who must provide the initial emergency response. This training must provide participants with knowledge sufficient to recognize hazardous materials, know the risks associated with the material, and know how to respond to the discovery. First responders at an "operations level," who are those charged with protecting property, persons, and the environment, but not with actually handling contaminated materials, must have 8 hours of training in addition to the training provided at the awareness level.

5. Medical surveillance is required where respirators are worn or employees are exposed to hazardous substances at or above allowed exposure levels, though work procedures and approaches must be developed to maintain actual exposure levels below established exposure levels.
6. In some instances, air monitoring is required.
7. An information program must be developed and implemented to identify key personnel, especially those responsible for site safety and health.
8. An emergency response plan is required to address personnel responsibilities, places of refuge, security, evacuation routes, emergency medical facilities, and procedures for alerting and working with local emergency services.

In summary, once a hazardous material is discovered and cleanup operations begin, there are a number of basic regulatory requirements related to health and safety that must be met. The basic thrust of these regulations is to force those involved in projects where contamination exists or is likely to be found to evaluate the associated risks, to develop a plan for addressing these risks, and then to implement this plan. The focus is also on training employees to recognize hazardous materials, understand the risks, and respond appropriately.

Containment

Once contamination is discovered, it is important to limit its spread. The response may be as simple as closing and posting a door or as elaborate as evacuating endangered personnel and restricting access to the area. Ultimately, some effort may have to be made to halt or otherwise curtail the flow of the contaminant into the surrounding environment.

The first step to containing almost any contamination is to stop doing whatever released or exposed the contamination in the first place. The next step taken depends on the nature of the release and will vary from situation to situation. For example, if the contaminant is leaking from a pipe that is broken during excavation, the immediate action is to stop excavating. If, as in the project described at the beginning of this chapter, the contaminant is oozing slowly from the ruptured pipe, there may be time to notify the appropriate supervisory personnel before any further actions other than stopping work are taken. If the contaminant is spewing from an active ruptured line, however, then the next step may be to find the shutoff valve, or to evacuate the area. Whatever approach is taken to the containment of contamination, the likelihood that it will be accomplished successfully will be a function of the knowledge and experience of the project team. Thus, researching the project to discover likely hazards, hiring personnel with some experience with these hazards, and training them to deal with contaminants once they are discovered will enhance the chances that the project team will act promptly and correctly to contain any contaminants discovered on the site.

Keep in mind that responding successfully to the discovery of contamination may require project personnel to be aware of the limitations of their capabilities. In other words, when is it smarter to drop everything and seek help? The specific answer to this question will depend on the situation. Consequently, the chances that this question will be answered correctly by the project team will be enhanced by having project personnel who are experienced and well trained.

Evaluation

In the example described at the beginning of this chapter, it was unknown at the time of discovery whether the yellow ooze was hazardous or not. The project team properly assumed that it was, but for the project to move forward, their fears had to be either confirmed or put to rest. The specific steps in the process used to evaluate the potential contaminant depend on the nature of the contamination. In a general sense, however, evaluation proceeds much the same way regardless of the nature of the contaminant.

The typical evaluation process begins with identification of the person or persons who have the appropriate expertise to conduct the evaluation. Such expertise may be readily available on site. If it is not, then the project team may have to seek assistance from other sources. In the case of the yellow ooze, nobody on site could identify the potential contaminant. Consequently, a testing laboratory was located and hired to sample the potential contaminant and determine its composition. The chief problem with using an outside testing laboratory is the time required to locate a lab with the appropriate qualifications, make appropriate contractual arrangements, schedule the taking of a sample, and then the time required actually to test the sample and evaluate the results. If the project sits idle while this evaluation takes place, the cost of the resulting delay can be substantial.

The best way to avoid such a delay is to have identified the potential contaminant before it is discovered, or to have identified, selected, and made some contractual arrangements with a testing facility before the project is begun. For example, consider the yellow ooze. Some research into the history of the site would have revealed that a metal plating facility used to occupy the site. Consultation with people knowledgeable about such facilities might have revealed the types of contaminants that might be expected and their most likely location. If the presence of chromium wastes was a possibility, then a qualified testing lab capable of evaluating the presence of this contaminant could have been identified and contracted with before the contaminant was actually discovered. As it was, the public agency responsible for the project required a substantial amount of time to identify the testing agency and get it to the site so evaluation could begin. This time could have been saved (and a substantial amount of money as well) with a little research and planning.

Documentation

There are those who would argue that if it was not written down, it never happened. While some may view this as an overly legalistic view of the world, their perspective may perhaps be a little naive. There are reasons, however, other than the threat of litigation, to maintain adequate documentation. For example, recording the location where asbestos was discovered in a large hospital renovation project allowed project personnel to locate the asbestos for abatement once the abatement contractor had been mobilized to the site. Measuring and recording the amount of asbestos removed allowed the parties to determine the appropriate payment for the abatement work once it was completed. Recording the time the project crews sat idle while the abatement work was performed allowed the project team to measure accurately the cost and time impacts of the discovery of the asbestos. On a large, complicated project, particularly a project that has encountered problems in the form of unexpected asbestos or some other contamination, it is not realistic to think that project personnel will be able to retain every important fact in their heads. Hence, the need for adequate documentation.

The line between adequate and inadequate documentation may not be a clear one. As a consequence, it often pays to be conservative and overdocument rather than underdocument. The events and facts that need to be documented may also not be obvious beforehand. In general, it is important to consider why documentation is being maintained and then to document the salient facts and events that conform to these needs.

Several types of documentation are standard on most construction projects. These include daily reports or logs, diaries, telephone logs, meeting minutes, letters, schedules and schedule updates, pay requests and other invoicing documentation, submittals, foremen's reports, memoranda, change orders, job cost reports, and the contract documents. There may also be specific forms or types of documentation required by the contract. In addition to these standard documents, there may be several forms of documentation peculiar to a particular owner or contractor, or type of project. In addition, the standard forms of documentation may go by different names on different projects. Each type of document could potentially play a role in the documentation of the discovery of contamination on the project site.

Daily reports, logs, and diaries

At least one of the parties on a construction site will often take the initiative or have the requirement to prepare some kind of daily record of project events and other data. Figure 5.2 is an example of such a daily record. Note that a separate log will be maintained for each contractor or subcontractor on the site. Note also the reference to the back side of the log for the recording of overflow or additional information. A second page could also be used. Many owners, particularly public owners, have standard daily report forms. In the event that contamination is discovered on site, the daily report on the day the contamination was discovered should contain all of the basic information related to the discovery itself. This would include the time and location of the discovery, the personnel involved in the discovery, and the immediate actions taken following the discovery. The daily log can also be used to indicate the personnel and equipment idled by the discovery. For example, an "X" next to the backhoe entry can be used to indicate that the backhoe was idled by the discovery of the contamination. An "X" next to the operator labor category would be needed as well. On subsequent dates, the daily log would be used to continue to record the crews idled and the work that was not performed due to the discovery of the contamination.

If the information related to recording the discovery of contamination was particularly voluminous or there were numerous discoveries during the duration of the project, consideration might be given to preparing a separate log devoted solely to recording information associated with the discovery of the contamination. These special log sheets might be typed, printed, or copied onto colored sheets of paper so that they can be easily distinguished from the regular project logs.

Remember that any document prepared for the project, even for personal use, is a *discoverable document*. This means that in the event of litigation, all involved parties will most likely have a right to have a copy of the daily logs, diaries, and other documentation if this documentation contains information relative to the project and the issues in dispute. So do not write things in a log or diary that might be regretted later.

The best advice when preparing any project documentation, including daily logs, letters, memoranda, or meeting minutes, is to be professional and stick to the facts. Do not write things in such a way that they might later be interpreted as biased or unprofessional. Do not use coarse language or be flippant or satirical. Be descriptive and complete, but keep it simple and clear. Avoid words and phrases that are overly colorful or needlessly antagonistic. Handwritten logs are perfectly acceptable and normal, but when handwritten information is not legible, consider typing.

Letters and memoranda

While daily records are useful for recording the details associated with the discovery of contamination on site, they are inadequate for providing written communication. This is because daily logs typically record only raw facts and

DAILY REPORT

PROJECT: _____

Day _____ Date _____ Weather _____
Temperature (°F) Min ___ Max ___ Tide Conditions / River Elevs _____
Contractor/Subcontractor _____ Crew _____

Description & Quantity of Work Performed

Labor	No.	Hrs.	Equipment	No.	Hrs.	Equipment	No.	Hrs.	Equipment	No.	Hrs.
Supt.			Air Compressor			Crane, Truck			Rotary Tiller/Mixer		
Foreman			Air Hammer			Ditch Witch			Semi-Dump		
Surveyor			Asphalt Distributor			Earth Mover			Trenching Machine		
Operator			Asphalt Paver			Frontend Loader			Truck, Auger		
Driver			Backhoe			Generator			Truck, Dump		
Flagman			Bobcat Loader			Mechanical Tamp			Truck, Flatbed		
Semi - Skilled Laborer			Bulldozer			Milling Machine			Truck, Grease		
Common Laborer			Concrete Bucket			Mini Roller			Truck, Lane Striping		
Administrator			Concrete Saw			Motor Grader			Truck, Pickup		
Office Engineer			Concrete Screed			Power Broom			Truck, Pipe		
Broom Man			Concrete Vibrator			Pump			Truck, Water		
Carpenter			Crane, Grove AT			Roller, Rubber Wheel			Van		
						Roller, Steel Wheel			Well Drilling Rig		

Materials: (note whether received or used) | Visitors (Name - Representing)

Use other side as needed for additional explanations, sketches, etc.
Note: I=Idle, Not Used

Signature

Figure 5.2 Daily report.

data. They do not explain why the information was recorded, and they do not tie together and explain the significance of different facts on the same log or from day to day. When the significance of information must be explained or the results of analysis communicated, a letter or memorandum is best suited for this. In fact, whenever it is necessary to communicate anything other than the most basic information to another party on the project, a letter is appropriate. Though letters cannot substitute for face-to-face communication, they are essential to such tasks as providing notice, documenting the position on a particular issue, and answering the written positions, arguments, or questions proffered by other parties to the contract. With regard to this last item, it is easy to get into a letter-writing war. Resist the temptation. Letters do not build projects, and it is not true that the most prolific writer wins every dispute. While it is probably necessary to respond in writing to each written assertion made or position taken by other parties, it is not necessary to have the last word and there is no real need to repeat your arguments over and over, in letter after letter.

Keep letters simple and as short as possible. Try to address only one issue in each letter. When referring to documents or earlier letters, where it is practical, try to include the referenced documents as exhibits or quote from the referenced documents in the text of the letter.

Contractor, owners, and others on the construction site should not shy away from stating and arguing their position on an issue in writing. It is often remarkable how helpful writing a position down is to focusing thoughts and improving the logic and clarity of the position. Look at any letter written to convey a position as a sales document. It should be a persuasive statement of the arguments and logic on which a position is based, with the objective of convincing the person who reads the letter that the position is correct. Try to anticipate all facets of an issue and address them even before someone else raises them.

Meeting minutes and other project documentation

The recommendations concerning the preparation of meeting minutes and other project documents that relate to the discovery of contamination are essentially the same as those related to the preparation of daily logs and letters. It is important to stick to the facts, and to present information clearly and professionally. With regard to the preparation of meeting minutes, there are several techniques that can make them more useful. First, the preparation of the minutes is a task that should be assigned to one party on the project, or maybe even one person. Minutes are most useful when they are used to track the resolution of specific issues as these issues arise during the course of the project. This is often accomplished by incorporating an "action items" list or log into the minutes. An *action items list* identifies each issue raised during the meeting that is not resolved during the meeting. It identifies the party responsible for addressing the issue or resolving the problem. Usually, the list identifies a due date or date by which the issue must be resolved or addressed. For issues that remain from one meeting to the next, the action items log will indicate the difficulties encountered and the additional actions required. By using an action items log, issues are not lost from one meeting to the next, and some form of accountability is maintained. Often, it is effective to identify action items with unique numbers. Any action item that is not resolved in successive meetings is retained in the minutes with the same number designation. In this manner, it can be determined when an action item is finally resolved, and all parties will be reminded regularly of any unresolved issues.

Meeting minutes also typically include a listing of attendees and a description of any discussions held, direction given, or problems raised and resolved. If meetings are tape recorded, it must be with the full knowledge of all involved. Recording meetings may be a bad idea, however, as it can hamper open and honest communication among the parties.

If meeting minutes are received which are believed to be inaccurate and not representative of the discussions that took place, it is generally a good idea to identify and describe the inaccuracies in a letter to the drafter of the minutes. Amendments can be noted in the next meeting's minutes.

Document organization

It is pointless to maintain a written record if this written record cannot be retrieved. Consequently, written documentation requires some method of organization for easy and quick retrieval. While there are numerous electronic document organization and filing systems, most projects still maintain extensive paper files. Organizing these files is generally more tedious than difficult. It is also where tenacity and good habits ultimately pay off. Some organizations have standardized their filing systems from project to project. Others have developed systems that work for them through a trial-and-error process over several projects and years of experience. In general, documents of a type are usually filed together. For example, daily logs are filed together in chronological order. The same generally holds true for meeting minutes. Change orders and other numbered documents are generally filed in numerical order.

The most difficult category of documents is usually letters and memoranda. For example, a letter from a subcontractor to a general contractor informing the general contractor of possible site contamination might reasonably be filed in a number of locations. It might be filed in a chronological, incoming correspondence file. It might be filed in a file set up especially for the subcontractor providing the notice. It might also be filed in a special issue file set up to track and gather all the documents related to the discovery of the contamination. Some contractors or owners may also file letters by specification section. And this is only for the project's general files. Separate files may also be maintained by the general contractor's subcontractor coordinator responsible for the subcontractor who provided the notice. In fact, to facilitate easy reference, the letter may need to be filed in all these places. That is one reason why it is generally a good idea to restrict each letter to one issue. If more than one topic is addressed in a letter, filing the letter correctly is more difficult.

Given the difficulty of filing important correspondence, there are several practices that can be adopted to facilitate filing. The first of these is the proper use of the subject or "RE" line. It is good practice to describe the issue addressed by the letter. For example, the subject line of a letter addressing the discovery of asbestos might read, "Discovery of Asbestos, Room 316." In addition, where more than one project may be ongoing on the site, a contract or project number may also be useful on the subject line. On projects with many issues it may be a good idea to assign an issue a particular number. For example, every time a contractor identifies an issue that it feels may affect the cost of performing the project work, it may establish a cost code number to track the time and expenditures associated with the issue. This cost code number could also serve as the contractor's identification code number for the issue. Every piece of correspondence associated with the issue is coded to the appropriate cost code file. By this means the contractor automatically tracks both the costs and the justification for a potential change order.

Similarly, the owner could also assign an issue number to problems as they arise. The owner might want the issue number to correspond to the number that appeared in the action items control log. One thing to avoid, however, is a proliferation of numbering systems, each describing the same issue. Thus, if everyone on the project wants to establish independent numbering systems to track issues, the parties should meet to discuss a common system to track issues. The action item contract log number is a natural number to use for this purpose.

Other forms of documentation

The forms of documentation of contamination on a construction project are not limited to letters and other written forms of communication. As discussed in the previous paragraphs, audio tapes can also be used to record events, though the use of audio tapes can be problematic. Events can also be photographed or videotaped. Neither photographs nor videotapes can substitute entirely for written documentation, but instead should be looked on as extremely useful supplements.

Photographs help those involved to understand and visualize a problem. They can be essential tools in the communication of the magnitude or specifics of particular problems to those not actually on the site, such as upper management, governing boards, designers, and others. Several types of photographs are generally taken on construction projects. Event-specific photographs are used to record a particular occurrence or situation. These are usually taken on an as-needed basis. For example, an event-specific photograph may have been taken when the yellow ooze was first discovered as described at the beginning of this chapter.

A second type of photograph is taken to record progress. For example, four cameras were mounted on poles or adjacent buildings during the construction of a convention center. The cameras were automatic and programmed to photograph the site once each day or more often as required. The purpose of these photographs was to maintain a visual record of the daily progress of the work. On particularly large or complex jobs, such a record can be very useful. It might also facilitate the management of a project where the project manager cannot be on site each day. Aerial photographs of projects spread over a large area can also be a useful way of tracking and recording progress.

When taking an event-specific photograph, it may be a good idea to use a Polaroid or similar camera. If you do not get a good shot, you will know it before it is too late to get another. Many people, however, prefer the picture quality of a 35-mm camera. Try to make sure that the proper scale is maintained in any shot. Place a hard hat or tape measure into the picture if some sense of scale is required, or have a person stand in the picture. Always make sure every picture is dated (buy a camera with a date diode), and that a written description of the location, vantage point, and contents of every picture is provided. Pictures are almost useless without some authenticating or descriptive information. It may also be useful to record the name of the photographer together with the picture.

Videotaping a project is a commonly used approach to recording significant project events. In certain situations it might even pay to use a professional videotaping service. A videotape is not a substitute for a daily log. It is too difficult to reference information in a videotape to allow it to supplant the need for a detailed daily log. Also, be careful of the audio portion of the taping process. While it can be useful to provide a narration for the images as they are recorded, a casual remark could be misconstrued, or inappropriate or unprofessional language might be picked up.

Videotaping is also an excellent way of recording the state of a large project at a particular point in time. For example, the cleanup of a former chemical weapons and insecticide manufacturing facility required that large volumes of material be trucked over local roads to a disposal site. The contract documents required the contractor to restore local roads after construction was completed to the condition these roads were in before the contract work was started. The local roads were not in very good condition prior to the start of work. The contractor videotaped the condition of the roads prior to the start of work. At the end of the project, the owner directed the contractor to repair the roads. The contractor then produced the videotape to show that the damage it was being asked to repair had existed on the site before work commenced.

Summary

Immediate responses to the discovery of contamination consist of discovery, notification, safety and health prophylaxis, containment, evaluation, and documentation. The chances of discovering and containing contamination before it has spread throughout the project or even outside the project's boundaries is significantly improved by a thorough evaluation of the project and the project site, and careful selection and training of project personnel. Effective notice is often both a contract requirement and essential to communicating the existence and extent of the problem to the project team. Once discovered, valuable time can be saved by quick and accurate evaluation of the contamination and actions to be taken to effect its containment and/or cleanup. A paramount consideration is tending to the safety and health factors for workers or any other individuals subject to exposure. For projects where the discovery of contamination is expected, time and money can be saved if the actions to take when the contamination is discovered are considered in advance. Finally, a fair and equitable resolution of the problem caused can be assisted by proper documentation of project events.

Remember also that the discovery and cleanup of contamination on a project site will trigger numerous regulatory requirements. With regard to worker health and safety, the primary regulations of concern are those promulgated by OSHA. The approach mandated by OSHA is essentially the same as has been described elsewhere in this book, but with a focus on proper documentation. OSHA regulations require research and planning. Only then can an effective management plan be developed.

Chapter 6

Measuring the Time Impact of the Discovery of Contamination

Having dealt with the discovery of contamination itself, the next step is to evaluate the impact the discovery may have on the project. Impacts can take many forms. At worst, the discovery of contamination might make a project too expensive to build and result in its cancellation. Or the project, or part of the project, might simply be shut down until the contamination can be cleaned up or removed, resulting in delays to some of the project work or to the project as a whole. It is this impact that is the subject of this chapter—the time impact of the discovery of contamination.

An event that impacts upon the time of performance of a work activity or project is called a *delay*. Delays are categorized by their impact on the anticipated project completion date. A delay that impacts on the project completion date is called a *critical delay* or a *controlling delay*. This comes from the concept that the project completion date can only be delayed along the critical or controlling path of work. As discussed in Chap. 3, the critical path of work is the path of work with the longest duration through a project. The term *controlling delay* is typically used on a project where a critical path method (CPM) schedule is not being used. The terms, however, are synonymous. Both critical and controlling delays are delays that would prevent a project from finishing as early as planned. For simplicity, the term *critical delay* will be used in this discussion.

Delays that do not affect a project's ultimate planned completion date are noncritical or noncontrolling delays. A *noncritical* or *noncontrolling delay* is simply a delay to a particular work activity or section of the project that does not result in postponement of the project's ultimate planned completion date. The terms *noncritical* and *noncontrolling* are also synonymous. The term *noncritical* will be used throughout the remainder of this chapter.

The methodologies used to calculate the magnitude of a critical or noncritical delay are the same, except for the last step. It is in this last step where the distinction between a critical and a noncritical delay is made. This distinction is important because the contractor's entitlement to a time extension and added compensation for the delay depend on the criticality of the delay. For example, a contractor is not typically entitled to an extension of the contract completion date for a noncritical delay.

The proper methodology to use in the analysis of a project delay depends on a number of factors, including the type of schedule being used to manage the project, whether the analysis is being performed before or after the delay occurred, and the detail and extent of the documentation available to conduct the analysis. Each of these factors will be addressed in the discussion that follows.

Analyzing Delays Using a CPM Schedule

From an analysis perspective, the easiest delays to quantify are those that occur on a project that is being managed using a CPM schedule. This is one reason, though by no means the only reason, that a CPM schedule is such a useful management tool. A CPM schedule allows a simple and accurate determination of the magnitude of a delay and the impact of the delay on other project work and the anticipated project completion date.

Before we describe the approach to delay analysis using a CPM schedule, a couple of rules need to be established. First, do not create a CPM schedule for a project for the purpose of analyzing delays once the delays have occurred. In other words, do not put together a CPM schedule after the fact, in order to analyze delays. Delay analyses based on CPM schedules created after the delay has occurred—or worse, after the project has been completed—are notoriously subjective. There is no one right or wrong way to develop a CPM schedule for a project, and it is just too difficult for an analyst creating the schedule after the fact to remain objective and keep natural biases, whether recognized or not, from creeping into the schedule. For this reason, the results of an analysis of delays based on a CPM schedule prepared after the fact almost always support the position that is most advantageous to the analyst. Though no one plans to end up with a project in litigation, it should be recognized that courts and other deliberative bodies have been highly skeptical of delay analyses based on CPM schedules created after the fact. On a more basic level, the development of an after-the-fact CPM schedule distracts the analyst and other parties to the dispute from focusing on the real issues. Instead, time is wasted developing the schedule and then arguing over its accuracy, reasonableness, and objectivity. The time would be better spent focusing on the dispute itself.

The second rule to follow when analyzing delays using the project's CPM schedule and schedule updates is to try to use the schedules actually used to manage the project. There will always be a temptation to modify the contemporaneous project schedules for the purposes of analysis. Do not succumb to this temptation easily. Unless the problems with the schedule or schedule update are or should have been obvious at the time the schedule was prepared and the errors identified affect the results of the analysis, it is probably unnecessary and generally inappropriate to modify the contemporaneous scheduling information. Also, if some adjustments to the contemporaneous information are necessary, be sure that these adjustments are based on information known at the time the schedule was prepared and not based on information that was only known after the fact. For example, an incorrect actual start or finish date may sometimes be assigned to an activity in a CPM schedule update. Though the daily logs prepared at the same time show that the work was started or finished on dates different from those indicated in the schedule, the error was not corrected. If it will make a significant difference in the results of the analysis, the error in the schedule can be corrected and the delay reanalyzed. Be careful, though: While adjustments to the actual start and finish dates of activities in a schedule or schedule update based on obvious errors may be appropriate, adjustments to the logic and planned durations of schedule activities should not be made unless the errors were recognized. Whenever schedule revisions are contemplated, keep in mind that some courts insist that any analysis of delays be based on contemporaneous and unaltered project schedules. The Veterans Administration Board of Contract Appeals, for example, has ruled that the parties to the contract will "live or die" by the contemporaneous scheduling documentation.

Other than these two basic rules, the analysis of time impacts associated with the discovery of contamination using project CPM schedules can be approached in a number of different ways depending on the circumstances encountered. The first and simplest approach is appropriate when the contamination is discovered before the affected contract work is performed. For example, asbestos is discovered unexpectedly in a renovation project. The asbestos is found in the insulation on pipes to be removed as part of the renovation work. Though abatement work had already been performed in the building, the abatement contractor apparently missed some of the pipe insulation. The rooms containing the pipe to be removed were not the first that would be renovated. Consequently, the owner had time to make arrangements with the contractor to have the asbestos removed before the contractor needed access to the rooms to do the renovation work. In this scenario there is sufficient time for the owner and the contractor to negotiate and agree to a change order modifying the contract to add the asbestos abatement work. Every change order should address both the issues of cost and time. Thus, the question of contract time should be addressed for each change. In this scenario the analysis required is similar to the process used to develop the project's CPM schedule in the first place. The first step is to become familiar with the additional work that must be performed. In this example, the added work consists of removing asbestos-covered pipe versus pipe from which the asbestos has already been removed. The next step is to develop a list of the added work activities that will be required. For this example, the added work includes contacting potential subcontractors to perform the work, since the general contractor is not experienced with such work or licensed to perform it; proposing the selected subcontractor to the owner for approval; awaiting receipt of the owner's

124 Chapter Six

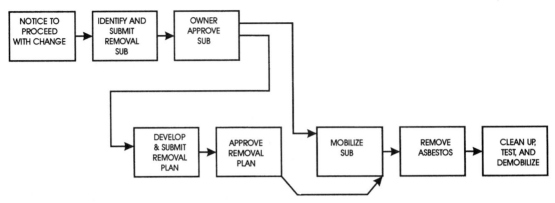

Figure 6.1 Fragnet for asbestos removal work.

approval; mobilizing the subcontractor after the appropriate review and approval of the subcontractor's removal methods; setting up a containment area for the performance of the removal work; actual performance of the removal work; and the follow-on cleanup and testing work.

Once the added work activities have been identified, the next step is to develop a fragnet. A *fragnet* is a CPM schedule logic network developed for a piece of a larger project. In this case, the fragnet describes the work associated with the additional asbestos abatement work. Figure 6.1 shows a fragnet for this work based on the previously developed list of activities. Note that the first activity in the fragnet is a notice-to-proceed (NTP) activity.

The fragnet also includes estimated durations for each of the listed work activities. As discussed in Chap. 3, the determination of durations for work activities is a function of the amount of work that must be performed and the resources available to complete the work. The planned duration for approval activities is often dictated by the contract. Unless it is already known that the added work will be performed on an expedited or accelerated basis, the durations used should reflect normal or typical durations for the anticipated quantity of work. After analyzing the impact of the extra work, it may be determined that an accelerated effort is required. Figure 6.2 shows the example fragnet with durations added.

Once a fragnet for the change has been developed, the next step is to add the fragnet to the existing project schedule or schedule update. The process used to add a fragnet to a CPM schedule is similar to the process used to develop the schedule in the first place. First, it must be decided how the fragnet will be incorporated into the existing schedule logic. Usually, this decision is a simple one, as the fragnet is inserted as a series of activities preceding the start of the work affected by the change. Figure 6.3 illustrates how the fragnet of Fig. 6.2 is incorporated into the existing schedule logic based on the scenario presented. The added asbestos removal work is simply added as work before the start of demolition work. The last activity in the fragnet is connected logically to the start of the demolition activity, which cannot proceed until the asbestos is removed. The first activity in the fragnet is not connected logi-

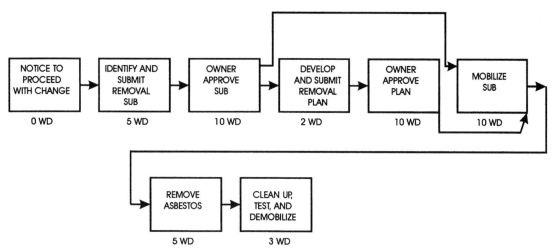

Figure 6.2 Fragnet for asbestos removal work with durations added.

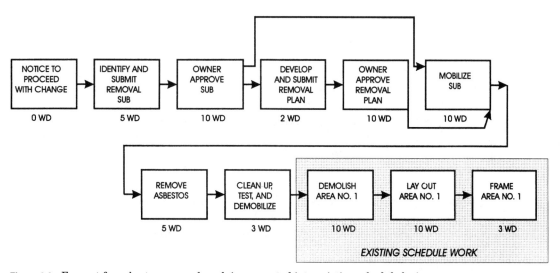

Figure 6.3 Fragnet for asbestos removal work incorporated into existing schedule logic.

cally into the schedule, since no schedule activities logically precede the change order activities. Note also that the existing schedule activities are numbered. The numbering of the change-order fragnet activities can either fit within the overall numbering scheme of the base schedule or can follow its own distinct numbering scheme. The use of a separate numbering scheme for change-order fragnets is probably the most common approach and allows the change-order fragnet activities to be clearly differentiated from the base schedule activities. The only problem with this approach is that when project activities are listed by activity identification number (in numerical order), the change-order fragnet activities will not be listed with the work activities

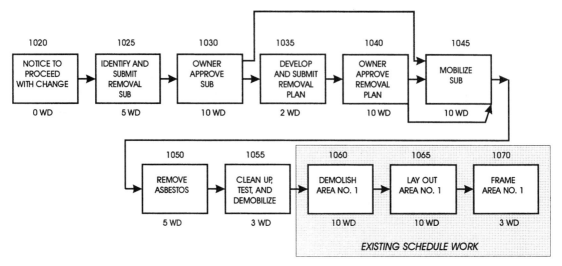

Figure 6.4 Fragnet for asbestos removal work with activities numbered.

affected by the change. If the number of change fragnets added to a schedule is extensive, it is probably best to number them separately. If there are only a few, incorporating them into the regular numbering system for the schedule may be preferable. In Fig. 6.4 the example fragnet activities are numbered using a sequence similar to that used for the base schedule activities.

Once the fragnet activities have been numbered, they can be entered into the schedule. The schedule or schedule update into which the fragnet is entered is usually the update in force at the time the fragnet is prepared, particularly if, as in this scenario, the contamination is discovered and corrective actions decided on before the affected contract work is scheduled to be performed. Having identified the appropriate update, the next step is to add the fragnet activities to the schedule. After adding the fragnet activities, the schedule is recalculated. If the addition of the fragnet activities results in a later project completion date than was indicated by the unaltered schedule or schedule update, then the change-order work will require that the project completion date be extended. The magnitude of the time extension required is the difference between the completion date projected by the unaltered update and the completion date of the update with the change-order fragnet added. Note that using this approach, the start date assigned to the fragnet NTP activity will be the data date of the schedule used. For example, if the data date of the schedule update used to analyze the example fragnet is May 1, 1996, then the earliest start date assigned to activity 1020, NTP with change, is May 1, 1996. Typically, however, as discussed in Chap. 4, the changes clause prevents the contractor from proceeding with extra work without written authorization or direction from the owner. Consequently, the date of NTP with a change may be different than the data date of the update used to evaluate the change. If it is significant (if it affects the duration of the time extension to be awarded), this problem can be minimized by restraining the start date of the NTP activity to the anticipated date of the owner's NTP with the change. If the owner gives direction sooner than anticipated, the duration of the time extension required can be reduced as appropriate. If the owner's NTP is delayed, the magnitude of the necessary time extension may have to be increased.

It should be noted that this brief analysis of the schedule is extremely important in the context of making management decisions on the project. For instance, if the example scenario shows that the abatement work will delay the project and require a time extension, the project manager might choose to evaluate whether acceleration is warranted for the work. For example, the work might be performed on a double- or triple-shift basis or with larger crews and more resources in order to minimize or even eliminate the delay. The premium to be paid to perform on an accelerated basis can be determined and a cost–benefit analysis performed to allow the project manager to make an informed decision as to how best to schedule the work and execute the change order. If the owner attempts to make decisions without the benefit of schedule analysis, the decision will be based on ignorance and may have serious cost and time impacts.

Analyzing Delays Using a Bar Chart

The analysis of the time impact of the discovery of contamination is easiest on a project with a well-maintained CPM schedule. On projects with bar chart schedules or no schedule at all the analysis process is similar, but the analysis is more difficult and more subjective. The first step, again, is to become familiar with the work to be added. Develop a bar chart schedule of the added work to be performed. A sample bar chart is shown in Fig. 6.5. This step is similar to the fragnet development step used when analyzing a CPM schedule.

The next step is also similar to the approach used when analyzing a CPM schedule, but because a bar chart is a purely graphical tool, all of the analysis work must be done by hand. First, redraw the bar chart to show the status of the project immediately before adding the change-order work. This step is similar to preparing a schedule update as of the date the change is to be added to the schedule. Next, add the bar chart for the change to the bar chart schedule for the project and redraw the schedule. If the redrawn schedule has a later completion date than the updated bar chart schedule, the contractor may be entitled to a time extension for the change. The magnitude of the time extension is the difference between the completion dates shown by the updated bar charts with and without the extra work added. This is done for our example in Figs. 6.5 and 6.6.

The analysis of a bar chart schedule is more difficult because it involves redrawing the project's bar chart. Note also that the results may be more subjective because bar charts typically lack the scheduling detail commonly found in a CPM schedule. As a result, there may not be an activity whose start is clearly affected by the addition of the change-order fragnet.

In summary, the process of determining the time impact of the discovery of

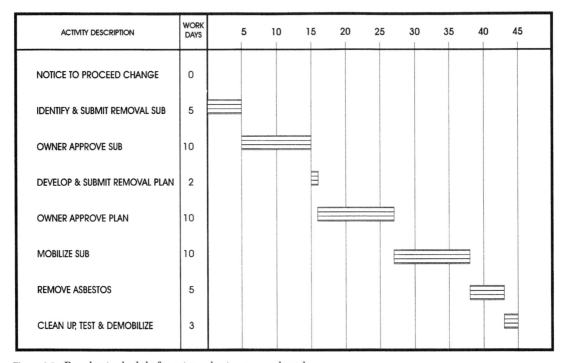

Figure 6.5 Bar chart schedule for extra asbestos removal work.

contamination before the corrective work is actually performed is a simple task. Become familiar with the change. Develop a fragnet or extra-work schedule for performing the added work. Add the fragnet or extra-work bar chart to the existing updated project schedule. The difference in completion dates between the altered and unaltered schedules is the measure of the time extension due the contractor for the added work, or an indicator that some form of acceleration should be considered.

Sometimes arguments are made that the time impact of a change cannot be calculated until after the added work is performed, or that the magnitude of a delay can be measured only in hindsight. The analysis procedure just described shows that the magnitude of a delay can be calculated before the extra work is performed. The time impact of a change can be estimated just as the cost impact of a change can be estimated. If a change order can be negotiated and signed on the basis of a cost estimate, then the time impact of the change can also be negotiated on the basis of an estimate. And just as it is desirable to come to some agreement on the cost of a change before the extra work is actually performed, it is generally beneficial to come to an agreement on the time extension due for the extra work before the work is performed. If the scope of the added work is known, and the need for the extra work is known timely, then the owner and the contractor should be able to estimate both the cost and the time associated with the change, conduct negotiations,

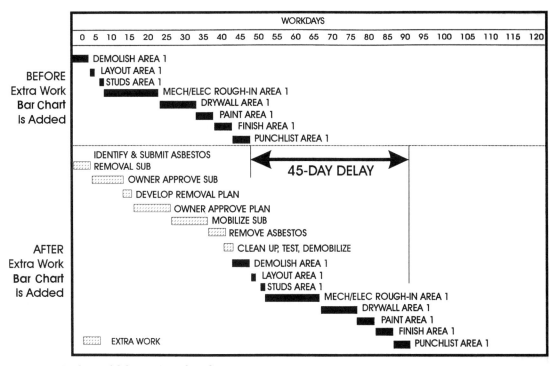

Figure 6.6 Analysis of delays using a bar chart.

agree to appropriate modifications to the contract cost and time, and execute a change order before the work is performed.

When the scope of the extra work is not known—a common problem when the extent of the contamination is not known—or the contamination is not discovered early, it may not be possible to negotiate a time extension before the work is performed. This problem may also result from a failure to recognize the extra work as a change until after the work is performed. In these situations, the magnitude of the delay may have to be determined in hindsight. Many analysts have developed different approaches to the analysis of delays after the delay has occurred. A discussion of these approaches and the strengths and weaknesses of each follows.

The Impacted As-Planned Approach

If the time impact analysis described earlier in this chapter is done after the extra work is performed, rather than before, and the impacted schedule is the original baseline or as-planned schedule for the project, then the analysis performed is known as an *impacted as-planned analysis*. The durations and logic used to develop the fragnets for the changes added to the schedule are based on the actual logic and duration of the extra work, not on planned logic and durations. Typically this analysis is performed by developing fragnets for a number of changes. These fragnets are then entered into the as-planned schedule and the schedule is recalculated. The difference in the completion dates between the original as-planned schedule and the impacted as-planned schedule supposedly represents the delay attributable to the added work. The strength of this approach is its simplicity: It is easy to do, and the results are easy to understand. As might be expected of an analysis approach that is so easy to perform, the primary weakness is a basic flaw in the logic of the analysis and the resulting inaccuracies in the results of the analysis.

The chief flaw of the impacted as-planned analysis approach is the assumption that the project was ultimately built in accordance with the logic and duration of the work as depicted in the as-planned schedule. It is the rare project that is built exactly as planned, especially when problems are encountered. Consequently, the assumption that the owner and contractor plow on, without deviating from the original plan, and make no attempt to minimize or mitigate the impacts of the problems encountered is a poor one. In addition, an impacted as-planned analysis is based on the assumption that the logic of the as-planned schedule cannot vary depending on the circumstances. True, some of the logic in any construction schedule is "hard logic," mandated by the physical realities of construction. For example, a bridge deck cannot be placed before the girders are erected. Demolition cannot be performed until the friable asbestos is removed. But much of the logic in a schedule may be preferential. For example, while the contractor or owner may prefer that the rooms in a renovation project be renovated in a particular order, there is often no physical reason why the rooms cannot be renovated in a different order, mitigating the delay that might be caused by the discovery of contamination in one of the first rooms to be renovated. Consequently, the assumption in the impacted as-planned analysis that the schedule logic is static and immutable is false.

The schedule for a project is dynamic. It changes over time based on the actual progress made, changes in the sequence of activities, additions of extra work activities, etc. Consequently, the critical path may change several times during the course of construction. The impacted as-planned analysis ignores this reality and hence is an academic exercise with no basis in reality. If the logic that underlies the impacted as-planned approach were accurate, then there would be no need to update the project schedule. Since it is obvious that schedule updates are useful and necessary, it is clear that the logic underlying the impacted as-planned approach is incorrect.

The inherent errors in the assumptions on which an impacted as-planned analysis is based are usually obvious when the results of such an analysis are reviewed. If the analysis fails to account for the efforts of the contractor and owner to adjust the schedule to mitigate the impact of problems encountered, then the analysis will tend to show more project delay than the project actually incurred, and will tend to show project activities starting and finishing later than they actually did. In other words, the best evidence of the errors in an impacted as-planned analysis is what actually happened.

Because of its inherent flaws and the ease with which these flaws can be illustrated, the impacted as-planned approach should never be used. The ease of use is overwhelmed by the flaws in the approach. Be particularly wary of an impacted as-planned analysis based on a CPM schedule prepared after the fact. As noted earlier in this chapter, CPM schedules prepared after the fact are notoriously subjective and often biased. Combined with the analytical flaws inherent in the impacted as-planned analysis, it is doubtful that the results would be a reliable measure of project delays.

Collapsed As-Built Analysis

The impacted as-planned approach adds fragnets representing alleged impacts to the project's as-planned schedule to measure the time impact. The *collapsed as-built approach* works in reverse. Impact fragnets are subtracted from an as-built CPM schedule that supposedly represents how the project was actually built. As with the impacted as-planned approach, the strength of this approach is in the simplicity of the analysis and presentation of the results. The weaknesses of this approach are also similar to those of the impacted as-planned approach. The collapsed as-built approach assumes that the project was planned to be built using the logic of the as-built schedule. Once again, it assumes that the critical path is static. A second problem with the collapsed as-built approach is the as-built CPM schedule itself. Such a schedule is usually created from the as-built information, including the project's daily logs and diaries, meeting minutes, and correspondence. This information describes when the project work activities actually took place and the duration of each activity. This information does not indicate, however, the relationships among these activities. This information must be supplied by the analyst. Unfortunately, as with any schedule created after the fact solely for the purpose of analyzing delays, information supplied by the analyst can be biased to favor a particular position. Note also that developing an as-built CPM schedule is a time-consuming, exacting, and expensive process. Given the weaknesses in the analysis, the result is not worth the expenditure of effort and money. Therefore, don't use it and don't accept it.

Contemporaneous Analysis Approach

Rather than rely on gimmicky analyses like the impacted as-planned or collapsed as-built approaches to perform an after-the-fact analysis of delays caused by the discovery of contamination on the construction site, use the project's existing scheduling information. On projects that have a CPM in place that was maintained throughout the project, the baseline (as-planned) project schedule and monthly updates should provide a more than adequate means of conducting a quick, unbiased, and accurate analysis. In some instances, courts and other jurisdictions have actually mandated that an analysis of project delays be based on the schedules actually used on the project.

Figure 6.7 represents the critical path of work for a bridge construction project. The critical path of work is the longest path of work through a proj-

ACTIVITY NO.	ACTIVITY DESCRIPTION	PLANNED DURATION (WD)	JULY	AUGUST	SEPTEMBER	OCTOBER
1	NOTICE TO PROCEED	1	1⏐1			
5	CLEAR & GRUB	5	2▭9			
10	EXCAVATE PIER 1	5	10▭16			
15	F.R.P. PIER 1 FTG.	10	17▭30			
20	F.R.P. PIER 1	15		31▭20		
25	SET GIRDERS	2			21▭22	
30	F.R.P. DECK	20			23▭27	
35	COMPLETE & OPEN	10				30▭11

Figure 6.7 As-planned schedule for a bridge construction project.

ect. Thus, it is the critical path of work that determines the planned project completion date. Only delays to the critical path of work can delay the anticipated completion date of the scheduled work.

Figure 6.8 shows how the contractor actually performed the critical contract work depicted in Fig. 6.7. This type of schedule, showing how the contractor actually completed the contract work, is often called an *as-built schedule*. An as-built schedule can be developed a number of ways. For example, using most scheduling software, a simple bar chart can be prepared using the actual activity start and finish dates recorded in each schedule update. While this approach is probably the quickest and easiest way to prepare an as-built schedule, an as-built schedule produced this way lacks important detail. Any interruptions in the performance of the contract work activities are not shown, since the length of the bar is determined only by the actual start and finish dates for the activity. The most detailed and often the most useful approach to creating an as-built schedule is to plot a bar chart using the project's daily logs and reports. Using this approach, the analyst will be able not only to identify and depict a work activity's actual start and finish dates, but any interruptions to the performance of the work as well. This added detail

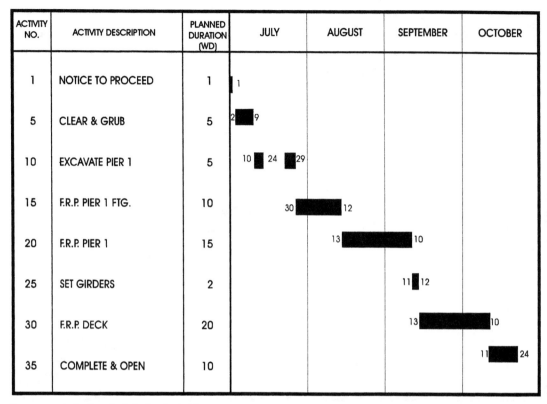

Figure 6.8 As-built schedule for the bridge construction project.

makes it easier to recognize and measure project delays. Plotting an as-built schedule from the project's daily records is accomplished much like any plot or graph. It can be done using standard, available computer software or by hand. The bottom or X axis is composed of a calendar. The left or Y axis is a list of the project work activities. If possible, this list of activities should be coordinated with the project's CPM schedule activities and should use activity names similar to those used by the CPM schedule.

Figure 6.9 is a plot of the as-planned and as-built schedules from Fig. 6.7 and Fig. 6.8 together. Using Figs. 6.7, 6.8, and 6.9, an analysis of project delays would proceed as follows.

The first critical activity is activity 1, Notice to Proceed. This activity has a 1-day planned duration and is expected to both start and finish on July 1. The planned start and finish dates for this activity are shown in Figs. 6.7 and 6.9. From Fig. 6.8 it can be seen that the notice-to-proceed activity was actually started and completed as planned on July 1.

The next critical activity in Fig. 6.7 is activity 5, Clear & Grub. The clearing and grubbing activity is planned to start on July 2 and finish on July 9. The as-built schedule in Fig. 6.8 shows that this activity was also completed

ACTIVITY NO.	ACTIVITY DESCRIPTION	PLANNED DURATION (WD)	JULY	AUGUST	SEPTEMBER	OCTOBER
1	NOTICE TO PROCEED	1	1 / 1			
5	CLEAR & GRUB	5	2—9 / 2—9			
10	EXCAVATE PIER 1	5	10—16 / 10, 24—29			
15	F.R.P. PIER 1 FTG.	10	17—	30 / 30—12		
20	F.R.P. PIER 1	15		3—20 / 13—10		
25	SET GIRDERS	2		21—22	11—12	
30	F.R.P. DECK	20			23—27 / 13—10	
35	COMPLETE & OPEN	10				30—11 / 11—24

Figure 6.9 As-planned versus as-built schedule for the bridge construction project.

as planned. No project delays resulted from the performance of this activity.

Given the July 9 completion of the clearing activity, the next critical activity, no. 10, Excavate Pier No. 1, should have started on July 10. This activity also started as planned. With a start date of July 10, this work should have been completed on July 16. As seen from the as-built data, this activity did not complete on July 16 as planned. Pier No. 1 excavation work was stopped after July 10 and was suspended from July 11 through July 23. Excavation work did not recommence until July 24. The period from July 11 through 23 represents a delay to activity 10 of 13 calendar days. Because the Pier No. 1 excavation activity is a critical activity, this delay is also a critical delay and represents a delay to the project completion date of 13 calendar days. To check this conclusion, calculate the revised completion date of activity 10. This activity has a 5-work day planned duration. One day of work was accomplished on July 10. Consequently, 4 days of work remain as of July 24. Given a July 24 start, these 4 days of work should end on July 29. (There is an intervening weekend. Since weekend days are not work days in this schedule, an activity with a 4-work day duration would require 6 calendar days to complete.) July 29 is 13 calendar days later than the original planned completion date for activity 10 of July 16. This is equal to the calculated delay, confirming that the calculated delay is

Measuring Time Impact of Discovery of Contamination   135

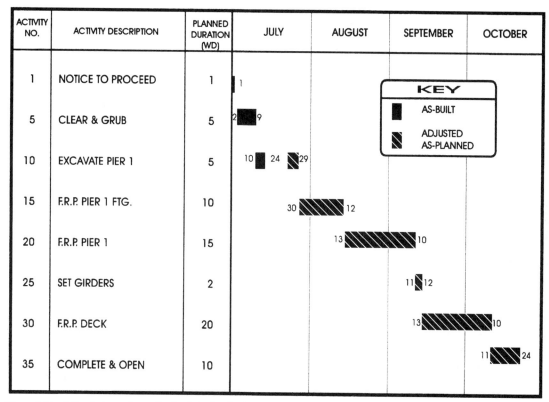

Figure 6.10 Adjusted as-planned schedule for the bridge construction project (as of July 24).

correct. A second check can be performed by adjusting the schedule of the remaining scheduled work for the 13-day delay. This schedule is presented in Fig. 6.10. This adjusted schedule is shown with the original as-planned schedule in Fig. 6.11. A comparison of the as-planned project completion date to the anticipated project completion date after adjusting the as-planned schedule for the 13-day delay to activity 10 shows that the planned project completion date slipped from October 11 to October 24, a delay of 13 days. This further confirms the accuracy of the original delay calculation.

Figure 6.12 is a comparison of the adjusted schedule with the as-built schedule in Fig. 6.8. From a comparison of these two schedules it can be seen that the remainder of the contract work was completed as planned, but 13 days later. Thus, the project was delayed a total of 13 days, and this delay was the result of the 13-day suspension of work on activity 10, Excavate Pier No. 1.

Having identified the source of the delay to the project, the next step is to determine the cause of the delay. For this example, the project correspondence and meeting minutes revealed that buried fuel tanks were discovered on July 10. Pier No. 1 excavation was halted until these tanks could be tested and removed. This work was not completed until July 23, and Pier No. 1 excavation work could not recommence until July 24. The hidden tanks were

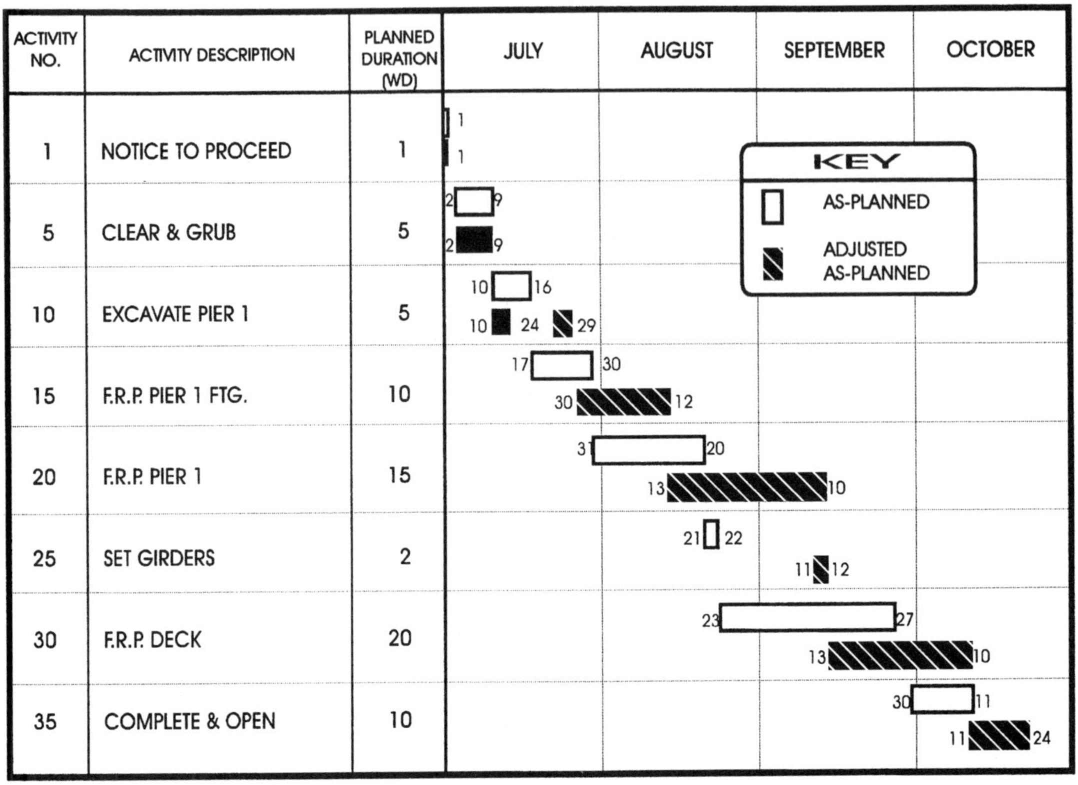

ACTIVITY NO.	ACTIVITY DESCRIPTION	PLANNED DURATION (WD)	JULY	AUGUST	SEPTEMBER	OCTOBER
1	NOTICE TO PROCEED	1	1			
5	CLEAR & GRUB	5	2–9			
10	EXCAVATE PIER 1	5	10–16 / 10–24–29			
15	F.R.P. PIER 1 FTG.	10	17	30 / 30–12		
20	F.R.P. PIER 1	15		3–20 / 13–10		
25	SET GIRDERS	2		21–22	11–12	
30	F.R.P. DECK	20			23–27 / 13–10	
35	COMPLETE & OPEN	10				30–11 / 11–24

KEY
☐ AS-PLANNED
▨ ADJUSTED AS-PLANNED

Figure 6.11 As-planned versus adjusted as-planned schedule for the bridge construction project (as of July 24).

the cause of the suspension of the Pier No. 1 excavation work and, hence, the 13-day delay to the project.

The previous example illustrates the approach to analyzing delays on a construction project after the delay has occurred. This kind of analysis is made much easier by the existence of an accurately prepared and updated CPM schedule. Using this kind of schedule, the project critical path is quickly and easily identified. It is also made easier by good project documentation. This documentation includes detailed daily logs or reports describing the work performed each day and the problems encountered. It also includes timely letters that provide a detailed description of the problems encountered, in this case the presence of hidden, buried fuel tanks, and detailed meeting minutes that record the actions taken by project personnel to resolve problems encountered.

The lack of a good project schedule and complete documentation of the problem does not prevent an analysis of delays from being performed. It will, however, make it more difficult to convince others, since the analysis will be more subjective and based on memory rather than letters, logs and diaries, and minutes of meetings. This is a good reason to maintain proper documentation on projects where the unexpected discovery of contamination is a possibility.

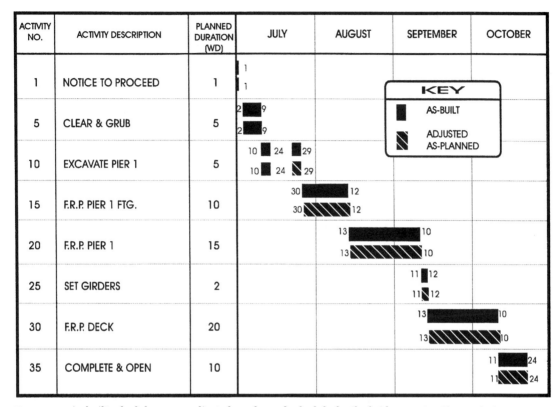

Figure 6.12 As-built schedule versus adjusted as-planned schedule for the bridge construction project.

Acceleration

It is appropriate to discuss delays and acceleration at the same time. Without one, there is often no need for the other, since it is a project delay that often necessitates acceleration of the project. The basic rule of acceleration is similar to the basic rule of delay analysis. Just as only delays to the critical path of work can delay the project completion, only by accelerating critical work can critical delays be mitigated and improvements in the project completion date be realized. Though such a conclusion seems obvious, there is often a tendency to accelerate everything, putting everyone on overtime, and extending working hours for all. Such a blanket and scatter-shot approach is often wasteful.

The sooner acceleration begins, the more likely it is to be accomplished at the lowest cost and with the least overall disruption to work on the project site. Thus, as soon as it is recognized that a project has been delayed, it is important to consider whether acceleration is warranted. In many cases, deciding whether to accelerate is a decision based purely on the least-cost alternative. For example, if an owner's liquidated damage figure, which supposedly represents a reasonable estimate of the daily cost to the owner of a delay, is $1000 per calendar day, then the cost to save a day of time should not exceed $1000 per calendar day. If the cost to save a day of time by accelerating exceeds $1000 per day, then it is more economical for the owner to let the project finish late. If, however, the cost to save a day is less than the amount the owner stands to lose if the project finishes late, then acceleration makes financial sense. There may be other reasons to accelerate or not to accelerate (the Olympic Stadium must be completed by the start of the Olympic Games, for example), but in most cases the decision to accelerate is based purely on economics.

There are three typical scenarios for acceleration. The first scenario exists when the project has incurred an excusable delay for which the contractor is entitled to a time extension and the owner desires to accelerate the project to avoid the delay. This is typically called a *directed acceleration,* because the owner directs the contractor to accelerate and reimburses the contractor for the costs incurred to accelerate. In reaching this decision, the owner often first asks the contractor to prepare a proposal identifying the critical activities that can be most economically accelerated and estimating the costs of the acceleration and the time to be saved. After discussing the contractor's approach and negotiating the costs and time, the direction to accelerate is normally accomplished by executing a change order. The change order both indicates the cost of the acceleration *and* moves up the contract completion date. In such a scenario, if the contractor fails to achieve the savings in the project duration, and the project still finishes late, the owner can reasonably assess liquidated damages for the late completion.

The second acceleration scenario is similar to the first, but the delay is not excusable. It is the contractor that is responsible for the time lost, and the contractor that must decide whether to accelerate. The contractor may have to consider more than simply an economic comparison of the liquidated damages that may be assessed versus the cost of acceleration. In some cases, failing to complete a project on time is considered a breach of contract for which a contractor may be terminated. If the alternative to acceleration is termination, the contractor's options may be severely limited. When accelerating work, the contractor must also evaluate proposals and issue change orders. The proposals, however, will be from the subcontractors, and the change orders will be written to the subs as well.

The third scenario occurs when the contractor incurs an excusable delay, but the owner refuses to provide a time extension. In such a case, the contractor must weigh the risks of being assessed liquidated damages or threatened with default against the cost of acceleration. If the contractor accelerates, this acceleration is often termed a *constructive acceleration.*

In order to establish that it has been constructively accelerated, a contractor must prove six basic points.

1. The project incurred an excusable delay for which the contractor was entitled to a time extension.
2. The contractor notified the owner of the delay and requested a time extension in accordance with the contract requirements.
3. The owner denied the contractor's request for a time extension.

4. The owner then expressed concern about the contractor's progress, directed the contractor to complete on time, or threatened the assessment of liquidated damages or termination if the contractor failed to complete on time.
5. The contractor accelerated.
6. The contractor's acceleration resulted in added costs.

If the contractor can establish these six points, then it may be able to establish entitlement to reimbursement of the added costs associated with the project's acceleration. The best way for the owner to avoid this kind of problem is to give the contractor time extensions when they are due. Don't wait until the end of the project. Since the unexpected discovery of contamination can often result in critical project delays, which are often excusable, the owner should consider the contractor's request for a time extension carefully. If a time extension is due, it should be provided promptly through a change order.

Chapter 7

The Impact of the Discovery of Contamination on Efficiency

The discovery of contamination on a project may affect the work in many different ways. Chapter 6 addressed the implications to the project schedule. The time impacts usually are easier to recognize and measure than other factors which are affected by the presence of contamination. One of the most significant areas to confirm is project efficiency.

A task that is accomplished efficiently is a task that uses the expected quantity or less than the expected quantity of resources to complete. For this reason, efficiency is usually measured in terms of the effort required to obtain a unit of output. For example, the efficiency of an excavation crew might be measured in terms of crew hours expended for each cubic yard of material excavated. These same units might also be used to describe the excavation crew's productivity. Thus, in common usage, the terms *productivity* and *efficiency* are synonymous. These two terms will be used interchangeably throughout this chapter.

The manager of a construction project must recognize that the presence of contamination can affect the productivity of operations on the job. By maintaining this awareness, the manager will be more likely to ensure that any change orders which are executed to handle the presence of contamination will include appropriate compensation for any adverse effects on the efficiency of the operations. With astute management and forward thinking, change orders can and should include all impacts. It is not desirable to leave the settlement of productivity problems to an after-the-fact time frame. These problems should be resolved contemporaneously with all related elements of a change order.

Contamination might affect the efficiency of the work in many ways. Some of these have been mentioned in other chapters. For example, the presence of contamination might cause a delay to the project. This delay might cause some of the work to be shifted in time so that it is performed in less favorable weather conditions. For instance, excavation operations originally planned during warm weather periods might be shifted into cold weather. The colder weather might adversely affect the productivity of the labor force and also the material characteristics of whatever is being excavated. A simple example is that the ground might freeze. Obviously, this will affect the ability to excavate the material. This obvious effect, however, might not be the only one. Not only may the frozen material be more difficult to excavate, the material might now not be acceptable to be used as backfill on the job. Consequently, the contractor might be required to import select fill in order to continue operations. Alternatively, the excavated material might have to be dried, scarified, etc., in order to make it acceptable to be used on the job. These types of effects are related both to productivity and to extra work. The overall operation of excavation and backfill is less productive, because extra work is required to meet the specification requirements.

Another effect associated with scheduling is that the contractor's sequence of work might be affected. Though the discovery of contamination might not delay the overall project, the contractor might have to alter the original planned sequence of work because of it. The contractor might shift crews to other parts of the job, or actually skip operations and return to them at a later time. Whether these actions affect productivity depends on the particulars of the project. It is quite likely that moving crews around will have some negative effect on the contractor's efficiency. There is some "down time" associated with moving out and back, sometimes referred to as mobilization and demobilization time. This is nonproductive time, but the cost associated with the labor and equipment hours is a real cost and may reasonably be charged against that specific item of work. The result is that the cost per unit of work achieved is higher. A similar effect associated with a sequence change is that the contractor's operations might be less efficient when the contractor is finally able to perform them. For example, if a contractor is forced to leave a portion of the work while contamination is cleaned up, then when the contractor returns, the area might be more congested, which could reduce productivity. Similarly, the ability to perform an operation efficiently might be adversely affected if the work is broken into smaller pieces during different periods of time. Though it is the same work unit, the piecemeal effort may yield a lower production rate.

The major consideration to keep in mind is that when contamination affects the project schedule in any way, the manager must carefully review every possible facet of the work that might be affected because of the change to the schedule.

Schedule considerations, however, are only one way in which efficiency might be affected. It is common for operations to be affected by the presence of contamination in ways totally unrelated to schedule considerations. A few examples should clarify this point.

For discussion purposes, assume that contamination has been discovered on the project site during the excavation for the foundations. The actual removal of the contaminated material should be treated as a separate and distinct work operation and should be handled by a change order. However, the original contract work might also be affected. For example, the contractor might continue to excavate the foundations around the contamination area. Because of the job-site layout, this might require that the contractor haul the excavated material over a longer route than originally planned. The net effect is that the cost per unit for the excavation is higher because of the additional hauling distance and time. In the same vein, access to the work area might be restricted by environmental operations associated with the cleanup work. This might lead to additional time or effort for base contract work. In some cases, workers working adjacent to a hazardous material might be required to wear protective clothing, might be required to wait while additional testing is done, might be interrupted during work by safety considerations, etc. All of these factors might reduce the productivity of the labor force in performing base contract work.

The effects to productivity on a job can be very broad and extremely difficult to recognize, especially before they occur. For example, on a highway project a rumor circulated that hazardous material existed in the area around a bridge abutment. Once the rumor "hit the street," workers refused to continue until the question was resolved. After careful testing, it was determined that no hazardous material was present. Despite this, workers were reluctant to work in that area, and the productivity on the project was measurably reduced. In this case, the mere hint of hazardous material caused a psychological response which affected production in a negative way.

If the original contractor for the project is also assigned the work of handling the contamination, the effort involved might distract the contractor's management staff from the base contract work. This is sometimes referred to as *dilution of supervision*. This is a very difficult area in which to measure variations in productivity. Therefore, any extra work should clearly define the amount of supervisory effort involved and, if necessary, separate management staff should be assigned to the extra work so as not to "dilute" the supervision of the base contract work.

Estimating the impact that a change might have on productivity, if any, is accomplished much like any estimating task. Cost and performance information is drawn from past experience, a clear and comprehensive understanding of the work involved, and published guides and estimating standards. The objective of a review of these sources is either to calculate the actual number of added labor or equipment hours required as a result of the anticipated inefficiency, or to develop a factor that can be used to adjust the estimate of the effort required to accomplish the work had no contamination been discovered.

Many projects have unit prices for specific items of work. This is particularly true for projects involving highways, site work, earth work, etc. Obviously, these projects are susceptible to impacts from contaminated soils. Often, it is presumed that the existing contract unit prices will remain valid even with the discovery of contamination. This may not be the case. For example, in a project involving a site development contract, contaminated material was discovered during excavation operations. In order to resolve the problem and maintain continuous operations to meet a rigid time schedule, the engineer revised the design of some of the features of the work. The engineer and the owner insisted that the contract unit prices would not be affected and applied them to the redesigned work being performed by the contractor. This situation ended in a lawsuit. What was not recognized was that the redesign altered the basic character of some of the units of work which were envisioned by the bidders under the original contract scope of work. Two examples will help clarify this.

One contract unit price was for rock excavation. The unit price was for a cubic yard of rock excavated and included all elements of the operation, such as drilling, presplitting, blasting, loading, and hauling. The contractor's bid was based on an analysis of the volume of rock it could excavate with each presplit blasting sequence. Based on that, the contractor was able to define a reasonable number of yards of rock produced for each cycle of the operation. When the project design was altered, the amount of rock which could be produced during a cycle was drastically reduced. Consequently, the costs were almost the same but the volume was markedly reduced. Thus, the contractor's unit price was too low. This situation was most obvious along a rock wall cut. When the area was realigned during the design modification, the amount of the rock wall that had to be cut was much more narrow than under the original contract. Therefore, instead of a cut of 6 feet of rock, the cut was only 3 feet. The volume was effectively reduced by one-half, yet the costs were not. In this instance, the basic premise on which the unit price had been developed had changed. Therefore, the unit price should not have been applied. Rather, a new unit price or a lump sum should have been negotiated during the design change instead of trying to impose the existing unit price on the contractor.

In a similar situation, when hazardous material was discovered during the earth work stage, the sequence of work was changed in order to allow remediation to take place. During the remediating, the contractor was directed to work "wherever possible." The contractor performed in accordance with this directive. The owner insisted that the existing contract unit prices still applied. This was not the case. Because of the change in sequence, the haul distance for excavation was increased significantly. Since the unit price for earth excavation included hauling, the original contract unit price was too low and should not have been applied.

If the impact of the discovery of contamination on productivity is not determined before the change is executed or a change order is not executed before the work is performed, there are several ways to calculate the impact of the discovery of contamination on inefficiency after the work is performed. Keep in mind as well that if the contractor believes that the discovery of contamination will have a negative effect on productivity, but does not include these costs in its proposal and does not reserve its rights to recover these costs after the work is performed, entitlement to such costs may be waived by execution of the change order.

The Measured Mile

The preferred method of calculating inefficiency after the fact is to perform a "measured mile" analysis. A measured mile analysis is a comparison of impacted and unimpacted work on the same project. The difference in the efficiencies of the unimpacted and impacted work is the measure of the inefficiency attributable to the impact. The strength of a measured mile analysis is that it eliminates many of the variables that can cloud the results of other approaches to the analysis of inefficiencies. These other variables include differences in projects, terrain, location, crew size and makeup, equipment selection, management, and subcontractors. By eliminating as many variables as possible, the measured mile method of measuring inefficiency allows the analyst to focus on the impact of the change alone on the contractor's productivity.

The first step in a measured mile analysis is to calculate the measured mile. The measured mile is the productivity obtained on an unimpacted portion of the work. For example, a contractor accepts a contract to install air-conditioning ductwork in two similar buildings. While the parties believed and the contract documents indicated that asbestos abatement had been accomplished in both buildings, the contractor discovered after starting the work, that only one building had been completely abated. Thus, while ductwork installation proceeded smoothly in one building, work in the other building was interrupted by the discovery of the asbestos and the coordination necessary to work in the building where possible while the asbestos was being abated. The measured mile in this case would be the ductwork installation productivity obtained in the uncontaminated building. This was calculated as 0.4 crew hour per pound of ductwork installed.

The next step is to calculate the productivity achieved in the impacted building—the building that had not been abated. In this building, the contractor was able to install at a rate of 0.5 crew hour per pound of ductwork installed.

Based on the productivities calculated for the measured mile and the impacted work, the presence of asbestos resulted in an inefficiency of 20 percent. This factor was calculated as follows:

$$\frac{0.5 \text{ crew hour per pound} - 0.4 \text{ crew hour per pound}}{0.5 \text{ crew hour per pound}} \times 100\% = 20\%$$

To find out how much time was lost to inefficiency in the impacted building, multiply the number of ductwork installation crew hours expended in the impacted building by the calculated inefficiency factor. If 500 crew hours were used in the impacted building, then 100 crew hours were lost to inefficiency. This figure was arrived at by multiplying 500 crew hours by the 20 percent inefficiency factor.

The most common reason why a measured mile analysis cannot be performed is that the project lacks a clearly identifiable measured mile period. Sometimes such a period is not easily identifiable because it simply does not exist. More commonly, however, a measured mile period cannot be found because the available project records are inadequate. This might be because the labor hours expended to perform a particular unit of work cannot be segregated, or because the number of units of work was not recorded. In situations where adequate production records have not been maintained, alternative measures of production can be considered. For example, most projects have a breakdown of work activities by cost, called a *schedule of values*. As work in a particular category is completed, the contractor is paid a percentage of the scheduled value based on the percentage of the work item completed. This information can form the basis for a very rough measure of productivity in terms of percentage of work performed per pay period. There are obvious weaknesses in this approach. The percentage of completion is a subjective measurement typically made by the contractor as part of the pay application process. Also, the period of time is so long (pay period to pay period) that it is difficult to measure the impact of changes that affect the project for only a short time. It is also possible that the percentage of completion may be influenced by the delivery and payment for materials, whether these materials are actually incorporated into the work or not. Consequently, inefficiency factors based on percentages of completion can be distorted by factors that are not related to productivity, and so this method is not recommended.

The best way to avoid the kinds of problems described is to maintain sufficient documentation to allow proper analysis to be performed. If the information is not recorded elsewhere, then good daily reports should record not only the kinds of work activities that were performed, but also the quantity of work performed and the labor and equipment hours expended to complete the work each day.

When the Measured Mile Cannot Be Used

In situations where a measured mile analysis on the impacted project cannot be performed, usually because there is no measured mile or unimpacted period against which to measure, the next best methodology is to perform a measured mile analysis by using a similar project. Obviously, the more similarities there are between the impacted project and the project productivity used for the measured mile, the more persuasive the analysis will be as a measure of the lost productivity.

When choosing a project for comparison, select a project that is similar in terms of the scope and quantity of work to be performed, using similar equipment, and a similar crew composition and size. An adjacent project using the same crews and equipment and performing identical work would be best. For every difference between the measured mile project and the impacted project, try to evaluate how the difference might affect the productivity calculations. For example, if the measured mile crew is using a different piece of equipment than the impacted crew, try to evaluate how the difference in equipment affects the productivity of the crew. If this question cannot be answered from the available data, consider contacting the manufacturer or others with experience with the equipment to try and discover how the difference might affect the productivity of the projects' crews.

Other than the fact that the analysis is based on different projects, the inefficiency factor calculations performed using separate projects are the same as those used when the measured mile calculation is based on the productivity achieved on impacted and unimpacted periods on the same project.

Absent a measured mile form of analysis, there are a number of approaches to the calculation of inefficiency. These approaches tend to be much more subjective and less convincing. One approach involves the use of an independent expert. This expert provides the measured mile productivity based on experience and knowledge of the impacted work. Problems with this approach include establishing the expert's objectivity and establishing the similarity of the expert's experience to the impacted work. This is also not a recommended approach.

A second approach might be to rely on inefficiency factors published by industry organizations. For example, the Mechanical Contractor's Association has published a listing of the inefficiency factors associated with various situations that might affect productivity. Be very careful when using these factors, and use them only as a last resort. These factors are often based on a limited amount of statistical data, if any at all. Because of this limited support and because they are usually published by contractor trade organizations, owners are often doubtful of the objectivity and accuracy of the published inefficiency factors. Consequently, rather than assisting in the calculation of inefficiency factors and the resolution of disputes among the parties to the contract, the use of published factors may simply add something more about which to argue. If, rather than calculating the actual inefficiency factors, the use of published inefficiency factors is considered, one source that may have relatively more merit are those available from the U.S. Army Corps of Engineers *Modification Impact Evaluation Guide*. The applicable sections of this guide are provided in App. F.

The Total Cost Approach

Another approach that is sometimes used to calculate inefficiency costs is the total cost method. The *total cost method* of calculating damages argues that the difference between the "total" actual cost of performing the work marked up for overhead and profit and the estimated cost of the work is the proper measure of the cost impact of work inefficiencies attributable to the owner's actions. Using this method, the contractor asserts that it is entitled to recovery from the owner of the cost overruns associated with performing the work. Essentially, this method converts a fixed-price contract to a cost-plus contract.

Though it is possible that a contractor's cost overruns may be representative of the impact on inefficiency, there are at least three reasons why the total cost overruns may not be the owner's responsibility.

1. The contractor's anticipated cost to perform the work may have been underestimated. One sign that work may have been underestimated is a significant difference between the amount bid by the low bidder requesting

added compensation and the other bidders for the work. If the low bidder left a significant amount of money "on the table" and its bid price was significantly below that of the other bidders, there is a clear suggestion that the contractor may have underbid the work. Verification of this observation would necessitate a careful review of the contractor's bid papers. Note that this would be an excellent reason for wanting a contractor's bid papers to be escrowed prior to the start of work.

2. A second possible reason for cost overruns other than those attributable to inefficiency might be that the contractor encountered problems in the execution of its work unrelated to the source of the alleged inefficiency. In other words, it requires the contractor's performance of the work to have been faultless. For example, due to the default of a subcontractor, the contractor may have been forced to perform the subcontracted work on its own at higher cost. The owner would not be responsible for these costs. As another example, a labor shortage may have forced a contractor to pay a premium for labor or forced the contractor to use less experienced labor or labor of lower caliber. These costs also would not be the owner's responsibility.

3. Finally, the contractor has some responsibility to establish that the costs actually incurred were reasonable.

These three issues can be difficult for a contractor to address. Also, courts are often less than eager to accept damages on a total cost basis. This reluctance is at least partially a result of the fact that no clear causal relationship between the owner's actions and the costs concerned is established. The contractor cannot point to any one cost and say that this cost was solely the result of inefficiency resulting from the owner's actions. Courts will also often require that the costs associated with the inefficiency attributed to the owner's actions must be impossible to segregate from the other project costs. In other words, the contractor must show that project documentation was not and could not have been sufficiently detailed to allow the contractor to discriminate between contract costs and the costs associated with inefficiency.

Note also that the total cost approach may shortchange the contractor. In situations where the contractor was able to beat its estimate and perform the work for less than planned, the resulting savings will be used in the total cost approach to offset added costs that may be the owner's responsibility.

Acceleration

In situations where the contractor accelerates the work, the assertion is often made that the acceleration caused the work to be inefficient. First, there is no direct correlation between accelerating work to perform it in less time and reduced efficiency or costs increases. Figure 7.1 shows a common representation of the cost of performing work versus the time to perform the work. There is an optimum relationship between the cost of performing the task and the time taken to perform the task. As the time taken to perform work increases, the cost of providing equipment and administrative support typically increases.

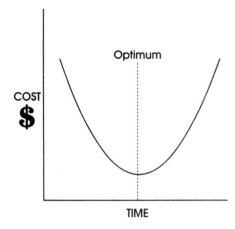

Figure 7.1 Graph of cost versus time.

As the work is compressed into shorter periods of time, however, the premiums paid and the impact of productivity losses that may result from long work hours, crowding, or night-shift work may become more pronounced. The optimum point is the point at which there is a balance between these two opposing forces. This relationship is illustrated by a factory working three shifts. The capital costs associated with constructing and maintaining the factory and its equipment offset the premium paid to swing- and night-shift employees.

With regard to cost increases attributable to acceleration, it is possible that if the starting point is to the right of the optimum point, accelerating the work will actually result in reducing costs. It should also be recognized, however, that working long periods of overtime can have a detrimental effect on productivity. Owners and contractors should evaluate these potentially detrimental productivity effects before embarking on long periods of overtime work. After consideration, it may turn out that adding crews and shifts may be a more cost-effective approach to accelerating a project.

In summary, the discovery of contamination on a project can affect the productivity of the contractor's work. Calculation of the impact of the discovery on productivity is best accomplished using a measured mile comparison of similar impacted and unimpacted work on the same project. If an unimpacted period cannot be identified, consider comparing similar work on similar projects. If a similar project is not available, an expert might be considered to help identify the measured mile productivity. As a last resort, look to factors published by industry trade organizations or the Corps of Engineers. Be careful of using the total cost approach to calculate acceleration costs. Also, be aware of the potential detrimental effects that working long periods of overtime can have on productivity.

Chapter 8

Measuring the Costs of Discovering Contamination

The discovery of contamination naturally brings up legitimate health and safety issues. Once these issues have been addressed, however, the focus of attention shifts quickly to costs. The question will almost always be the same: How will the cost of the project be affected by the discovery of contamination? Measuring the cost on the project is the focus of this chapter.

Estimated versus Actual Costs

As with delays and inefficiency, the approach to measuring the cost effect of the discovery of contamination depends on when the calculation is performed. If the contamination has been discovered but the remedial work has not yet been performed, the cost effect is determined by preparing a cost estimate. If the remedial work has already been performed, the cost impact is determined by identifying the actual costs. Since it is generally better to have a change to the contract in place before the extra work is performed, the preferred approach to calculation of the cost effect of the discovery of contamination is to prepare a cost estimate. Such an estimate is prepared much like any other and is based on experience, knowledge of the work, its impact on the project, and appropriate historic estimating information.

If the cost effect of the discovery of contamination is calculated after the problem has been corrected and the costs have already been incurred, the approach to the calculation of the cost effect is essentially a compilation of the costs incurred.

Whether the cost effects are estimated before the work is performed or accumulated after the work is completed, the basic categories of costs are the same. The discussion that follows assumes that the cost effects are being calculated after the work has been performed.

Labor Cost Effects

Labor cost effects are those costs that arise from increases in labor costs related directly to the discovery of the contamination. Labor cost effects themselves may fall into a number of different categories.

Additional labor

If the discovery of contamination adds work to a project and additional labor effort is required to accomplish this added work, then one cost effect of the discovery of contamination is the added cost of this additional labor. For example, the discovery of contamination often requires that a portion of the project be fenced or barricaded off from the rest of the project. If this fence would not otherwise have been required, then the labor expended to erect the fence is additional labor. The added labor costs associated with inefficiency fit into this category as well. The cost impact of this added labor could have as many as three components. The first component is the actual *direct cost* of the labor. This cost is calculated by multiplying the number of added hours of labor expended by the base hourly rate for field personnel. The number of hours actually expended is normally taken from the contractor's daily logs or foreman's reports. The time may also be recorded by the owner or its agent if the owner is monitoring the expenditure of effort on the added work independently. The hourly rate is available from the certified payroll records maintained by the contractor on the project. On some projects, hourly rates may also be stipulated by labor agreements. If salaried personnel perform some of the added work, then an hourly rate can be calculated by dividing the weekly salary amount by the number of hours worked each week.

The second component of the added labor cost is known as *payroll burden*. This category includes labor costs other than direct hourly wages or salary, such as health insurance premiums (if paid by the employer), the employer's share of FICA, the employer's share of unemployment insurance premiums, other insurance carried by the employer for employees, paid vacation and sick time, and other employee benefits and costs. The payroll burden associated with added labor expenditures is normally calculated by marking up the direct labor costs by some percentage. This percentage usually represents the combined cost of benefits as a fraction of the contractor's total direct labor cost. Note that this approach is really an approximation, since some payroll costs are not strictly a function of wages or salary. For example, the maximum amount of wages or salary on which FICA taxes can be assessed is capped. Once the cap is exceeded, no further tax is paid. In addition, medical benefits are not a function of wages and salary, but of family size, sex, and age. The payroll burden rate is usually known by the contractor, or can be calculated by the contractor's accountant. In some cases, the owner may want to calculate the payroll burden rate independently. For example, the federal government, particularly in cases where a large amount is in dispute, may conduct an audit of the contractor's financial records to verify the accuracy of the contractor's claimed amounts. The right to perform such an audit is provided for in the contract between the contractor and the federal government. A similar clause should be considered when the owner is drafting its contract for the project. The following is an example of such a clause.

> 109-15 CONTRACTOR'S COST RECORDS. The Contractor shall maintain records of all required payrolls, and of the details that comprise his total cost pursuant to any of the provisions under §104-03, Contingencies, Extra Work, Deductions, and he shall, at any time within 3 years following the date of final payment of the project, make such records available, on request therefor, to the Department for review and audit, if deemed necessary by the Commissioner. In case all or a part of such records are not made so available, the Contractor understands and agrees that any items not supported by reason of such unavailability of the records shall be disallowed, or if payment therefor has already been made, the Contractor shall, on demand in writing by the Commissioner, refund to the Department the amount so disallowed.

In situations where the added work is performed outside the normal working day, the third component of added labor costs is the premium paid for overtime or shift work. In most cases, this cost is paid only if agreed to by the owner in advance. In some cases a contractor may be limited to reimbursement for the premium time only by the contract. No payroll burden is allowed. This limitation reflects the fact that many benefits are not based on the amount of wages or salary paid. In this situation, however, a contractor may be undercompensated for the employer's contribution to FICA and other taxes and costs that are a function of the base labor cost.

Contracts also often contain provisions limiting the personnel that can be charged as direct labor expense. Many contracts preclude payments for the cost of management and supervisory personnel. The argument is that the cost of management and supervision is reimbursed via the markup for overhead allowed on direct labor costs. The contract should be reviewed carefully to identify any restrictions to reimbursement for additional labor costs.

Extended and idle labor costs

Sometimes additional labor expenses result not from the addition of work, but from delays to the work. These labor costs are often referred to as *extended labor costs*. A subset of extended labor costs is *idle labor costs*. Idle labor costs may result when interruptions or suspensions of the work resulting in crews sitting idle or underemployed until the problem that caused the work to be shut down is resolved. Extended or idle labor costs are calculated the same way as additional labor costs. The number of hours the employee's time is idled or extended by the delay is multiplied by the hourly rate from the certified payrolls. To this amount is added an appropriate markup for payroll burden. Typically, a suspension or interruption in the work should not result in premiums paid for overtime, since if the interruption occurred while the employee was working overtime, the employee would simply be sent home. If, however, the interruption or suspension occurred on an evening or night shift, some payment reflecting the shift premium might be due. Of course, if the contractor was given a directive to accelerate or constructively accelerated to make up lost time, some premium payment for overtime work might be due.

It should also be remembered that some kinds of delays are not reimbursable or compensable. Delays that are compensable should be described in the contract. If they are not, compensable delays are usually delays that were the fault or responsibility of the owner. These kinds of delays normally include delays resulting from design errors or changes, differing site conditions (if the contract contains a DSC clause), and, as a specific example, the unexpected discovery of contamination. Many other kinds of delay (strikes, hurricanes, material shortages) may be excusable, entitling the contractor to a time extension, but they may not be compensable—because these types of delay are neither the owner's nor the contractor's fault or responsibility. Consequently, the philosophy is that the owner accepts the additional cost of having a delay to the project and the contractor accepts the additional costs of being on the project for an extended period. The "no fault" approach ensures that neither side is penalized for occurrences beyond either party's control.

Labor escalation

One consequence of a project delay might be that the project work is postponed into a period of higher wages. As a consequence, labor costs are higher than they might otherwise have been. The calculation of the cost effect of wage rate escalation attributable to a project delay can present varying degrees of difficulty, depending on the scheduling and cost information that is available. In the simplest case, consider a project that was to finish in one construction season. No wage rate increases were anticipated during the period when project work was to be performed. Suppose, though, that as a result of the unexpected discovery of contamination on the site, work was suspended for several weeks. Because of the suspension and the associated project delay, the contract work could not be completed until the next year, after a $0.50 per hour wage increase. The contractor performed 1000 hours of work during the second construction season at the higher wage rate. What is the cost effect of labor escalation? The answer is $0.50 per hour times the 1000 hours of work performed at the higher rate. To this amount must be added an amount to cover the payroll burden. Note also that labor costs can escalate as a result of increases in benefit costs.

Calculations of the cost effect of labor cost escalation are more difficult on multiyear projects, where several wage rate increases are expected during the course of the project. This scenario would be further complicated by compensable delays stretching out for one, two, three, or more years. The guiding principle in these cases is to remember that labor cost escalation relates only to the labor hours expended after a wage rate increase that, but for the delay, would have been performed before the wage rate increase. For example, consider a project with an 18-month planned duration. A wage rate increase is expected

6 months into the project. An early delay to the project, the result of the unexpected discovery of asbestos in the floor tile to be removed during the renovation, resulted in the loss of a month. Some would look to the end of the project to see if any labor cost escalation occurred. Since there was no wage rate increase within a month of the end of the project, one would conclude that no escalation occurred. This conclusion would be incorrect. The proper point of reference is the date of the wage rate increase 6 months into the project. Was there work performed after that date that, but for the 1-month delay, would have been performed before the wage rate increase? If so, the number of added hours expended after the increase is subject to escalation. The calculation is performed the same way as the earlier, simpler example.

A few sample problems should more clearly illustrate the calculation of escalation.

Sample labor escalation problem 1. Figure 8.1 illustrates the anticipated labor distribution for a simple project. The contractor plans to mobilize a crew of eight for the entire 240-day duration of the project. At 120 days into the project, the wage rate for the project labor is scheduled by agreement to increase from $11.00 to $11.50 per crew hour. The laborers work an 8-hour day.

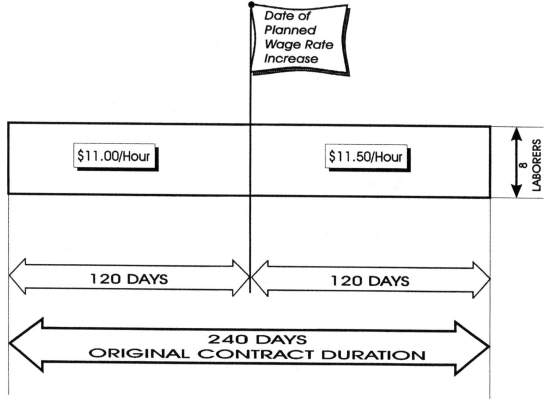

Figure 8.1 Anticipated labor distribution.

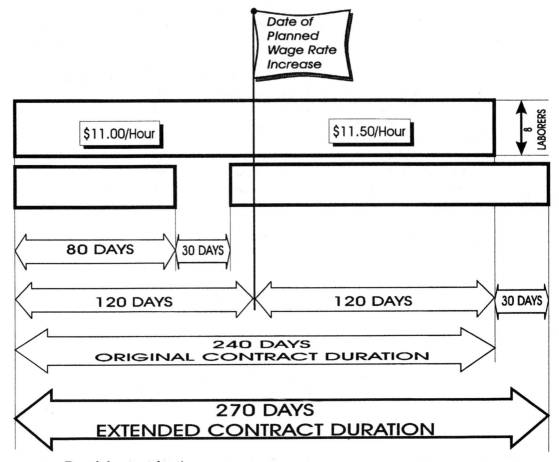

Figure 8.2 Extended contract duration.

As illustrated by Fig. 8.2, at 80 days into the project, asbestos is discovered in the building area undergoing renovation. Work is suspended for 30 days while the asbestos is removed. As a result of the work suspension, the project completion date is delayed 30 days.

The question is what, if anything, the contractor is due for labor cost escalation. The answer is $960. This figure is arrived at by the following calculation:

$$8 \text{ laborers} \times \frac{8 \text{ hours}}{\text{day}} \times 30 \text{ days} \times \frac{\$0.50}{\text{hour}} \text{ increase} = \$960 \text{ increase}$$

The basis for this answer is illustrated by Fig. 8.3. Because of the delay, 30 days of work that would have been performed before the wage increase were actually performed after the increase. Thus, the proper calculation of labor escalation would consist of 30 days of effort multiplied by $0.50 per hour for each hour of effort. There are a total of 1920 hours of effort over a 30-day period of the project. At $0.50 per hour, that equates to an escalation of $960.

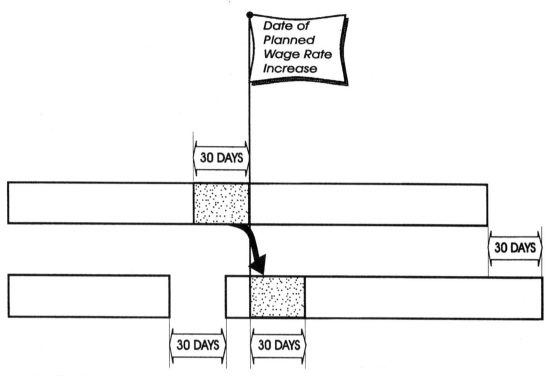

Figure 8.3 Sample labor escalation problem 1.

Sample labor escalation problem 2. In this problem, taken from a real project, the performance of drywall work was delayed by 71 days before the first wage rate increase on April 30. Table 8.1 shows the planned labor hour distribution for the project. Labor rate escalations occurred as of May 1, 1989, and May 1, 1990.

Numerous delays to the project occurred in addition to the 71 days, which were attributed to the owner's action. Along with the 71 days of delay, these other project delays affected the actual labor distribution. It was not possible to separate the effects of the intermingled delays. To estimate the effect of the 71-day delay, the planned labor distribution was first adjusted by 71 days. This adjustment is illustrated in Table 8.2. The results of this adjustment were then compared to the planned distribution to determine the escalation amount.

Table 8.3 presents a comparison of Tables 8.1 and 8.2. Based on the comparison in Table 8.3, the subcontractor is due $31,638 for labor escalation.

Equipment Cost Effects

Equipment cost effects arise from increased equipment costs attributable to the discovery of contamination on the site. These costs can be divided into three categories.

TABLE 8.1 Planned Labor Hour Distribution

Before 4/30/89						5/1/89 through 4/30/90												After 5/1/90	
12/88	1/89	2/89	3/89	4/89		5/89	6/89	7/89	8/89	9/89	10/89	11/89	12/89	1/90	2/90	3/90	4/90	5/90	6/90
5,805	5,805	5,805	5,805	5,805		5,805	5,805	5,805	1,451	0	0	0	4,354	5,805	5,805	5,805	5,805	5,805	2,907
Total = 29,025 Hours						Total = 46,440 Hours												8,712 Hours	

TABLE 8.2 Planned Labor Distribution Adjusted for 71-Day Delay

Before 4/30/89				5/1/89 through 4/30/90												After 5/1/90			
2/89	3/89	4/89		5/89	6/89	7/89	8/89	9/89	10/89	11/89	12/89	1/90	2/90	3/90	4/90	5/90	6/90	7/90	8/90
2,902	5,805	5,805		5,805	5,805	5,805	5,805	5,805	4,354	0	0	0	1,451	5,805	5,805	5,805	5,805	5,805	5,810
Total = 14,512 Hours				Total = 46,440 Hours												23,225 Hours			

TABLE 8.3 Escalation Calculation for 71-Day Delay

Escalation period	A As-planned hours	B As-planned with 71-day shift	C Escalation hours (B−A)	D Rate increase	E Escalation cost (C×D)
1. Start through April 30, 1989	29,025	14,512	0	0	0
2. May 1, 1989 through April 30, 1990	46,440	46,440	0	$1.52	0
3. May 1, 1990 through end	8,712	23,225	14,513	$2.18	$31,638
Total	84,177	84,177	14,513		$31,638

Additional equipment costs

Additional equipment costs arise from the addition of work to the contract. For example, if the discovery of contamination requires overexcavation, both the overexcavation and the placement of additional fill are additional work. Additional equipment costs are calculated by multiplying the number of hours of additional equipment operating time needed to accomplish the added work by the appropriate hourly rate. The number of hours of operating time may be available from the contractor's daily logs or foreman's reports, or from records maintained by the owner or its representative to monitor the performance of the added work. The appropriate hourly rate many times is defined by the contract and usually depends on whether the equipment is rented or owned.

Rented equipment rates. If the contractor has rented equipment specifically to perform the added work, reimbursement is simply the amount invoiced by the rental company and paid by the contractor. To this is added the cost of operation, mainly fuels and lubricants. If the equipment is delivered and picked up by the rental company, and this cost is included in the invoice and paid by the contractor, then no additional amount is due for mobilization and demobilization of the equipment. If not, then those costs should be calculated as well. If the cost of renting the equipment includes an operator, many construction contracts require the cost of the equipment and the operator to be separated, with the operator cost to be treated as a labor cost and the equipment cost to be treated as rented equipment.

If the contractor performs the added work with equipment that was rented to perform contract work as well, then an hourly rate must be developed for the rented equipment. This hourly rate is usually calculated by dividing the monthly rental amount invoiced and paid by 176 hours. If the rate is provided by the week or the day, then the rate is divided by 40 hours or 8 hours, respectively. Typically, the longer the equipment is rented, the lower is the effective hourly rate. The owner is typically entitled to the same rate paid by the contractor. Thus, if the piece of equipment is required for only one day to perform the added work, but the contractor has rented the equipment for the month to perform base contract work, the owner pays at the lower monthly rate, not the higher daily rate. To calculate the equipment cost effect, the hourly rate is multiplied by the number of hours the rented equipment is used to perform the added work. To this is added the cost of operation, again mostly fuel costs for rented equipment, though any other consumable not included in the rental rate is also included in the cost of operation. For example, filters and lubricants may be included as well.

Some equipment rental companies share ownership with contractors or are owned by contractors. In some of these situations the primary renter of the rental company's equipment is the contractor itself. A few owners have become concerned that the rates charged by such rental companies are not market rates. To assure that no more than the market rate is paid, these owners have drafted contracts that limit the amount paid for rented equipment to the market rate, or cap the rates paid at the rates listed in one of the equipment rate guides.

Owned equipment rates. If the equipment used to perform the added work is owned by the contractor, calculating an hourly rate can be difficult. This difficulty arises from the variety of methods used by contractors to track and apportion equipment costs. For example, one contractor whose records were recently audited did not accumulate or track costs by piece of equipment. Instead, all equipment costs were accumulated and tracked as a single item. Consequently, this contractor would find it very difficult to determine the hourly cost of a particular piece of equipment. Another contractor's audited records indicated that the contractor maintained accurate records of the cost of ownership for each piece of equipment owned, including maintenance and operating costs. To determine an hourly rate, the contractor divided the accumulated costs by the number of actual operating hours for each piece of equipment. This contractor had no difficulty coming up with an hourly rate, but the rates varied dramatically from month to month, even for identical pieces of equipment, depending on the number of hours of actual operation. Thus, even when a contractor keeps a careful accounting of costs by piece of equipment, it can be difficult to determine the basis for calculating an hourly rate.

Because of the difficulties associated with the calculation of an hourly rate for owned equipment, many contracts specify the use of one of several equipment rate guides available. One of the guides most commonly specified is the *Rental Rate Blue Book,* published by Dataquest. According to its introduction, the *Blue Book* provides rates "that an equipment owner should charge during rental or contractual periods to recover equipment-related costs on a single-shift basis." In other words, the *Blue Book* rates are calculated to represent the actual costs of equipment ownership. *Blue Book* rates are composed of four parts: depreciation, indirect equipment costs such as insurance and property taxes, cost of facilities capital (according to Dataquest this is not interest), and an allowance for major overhauls. In addition to the *Blue Book,* equipment costs are provided in the *AED Green Book,* also published by Dataquest. The *Green Book* is a nationally averaged compilation of equipment rental charges. Thus, this manual reflects not the costs of ownership, but the rental rates charged for equipment. Another source of equipment rates is the U.S. Army

Corps of Engineers' *Construction Equipment Ownership and Operating Expense Schedule.* This document is published in several volumes, with each volume covering a particular region of the country.

Not unexpectedly, the rates provided in the *Blue Book,* the *Green Book,* and the Corps of Engineers publication can vary significantly. The differences between the *Green Book* and the other two guides are attributable to the fact that the *Green Book* is merely a compilation of actual equipment rental rates, while the other two references are calculations of the cost of ownership. The differences in rates between the *Blue Book* rates and the rates presented by the Corps of Engineers appear to be based primarily on differences in accounting. For example, the *Blue Book* rates include the costs of insurance and property taxes. The Corps of Engineers does not include these costs in its rates. The Corps' position is that these indirect costs are covered by the contractor's overhead and profit markup. The Corps allows these costs, however, if the contractor's accounting system tracks these costs for specific pieces of equipment. The Corps also discounts the cost of facilities capital to reflect the portion of this cost that is recovered as part of the contractor's markup for overhead. The Corps uses the original purchase price of the piece of equipment for its depreciation calculations rather than the most recent manufacturer's list price. No discount is applied to the list price used by the *Blue Book,* although many pieces of equipment are sold at a discount off the list price. As a consequence of these differences, *Blue Book* rates are often higher than the corresponding rates allowed by the Corps' guide.

The *Blue Book* provides for a number of adjustments to the listed rates. Adjustments are described for equipment age, standby operation, job severity, duplication of costs, and discounts at time of purchase. Adjustments for climate and region are also provided as well as adjustments for multiple shifts.

In summary, the appropriate rate to use when calculating the cost effect of additional equipment use depends first on whether the equipment is owned or rented. If it is rented, reimbursement is usually at the rates invoiced by the rental company and paid by the contractor. If it is owned, the equipment rates used are as specified by the contract. If the contract is silent on this issue, the rate used might be based on the contractor's actual audited costs or on the rates provided in the guides, with appropriate adjustments. Whatever source is selected, when calculating the cost effect of added work, some amount should also be allowed for the operating cost of the equipment.

Extended or idle equipment costs

Extended or idle equipment cost effects arise from an interruption or suspension of the contractor's work. *Extended equipment* is equipment that is in operation longer than planned because of the interruption or suspension. For example, if illuminated signboards are used to warn traffic of a detour on a project and the discovery of buried fuel tanks causes the project to be suspended while the tanks are removed, then the added time the signboards are used is the extended equipment time. *Idle time* is the time a piece of equipment is actually idled or shut down by an interruption or suspension of the work. The cost effect of idled or extended equipment is calculated by multiplying the number of hours the equipment is extended or idled by the appropriate rate. The hours of extension or idleness are typically recorded in the contractor's daily logs or foreman's reports, or tracked by the owner or its representative.

The rate used in the cost effect calculation is again a function of whether or not the equipment is owned or rented. If it is rented, a rate derived from the amount invoiced by the rental agency and paid by the contractor is used. If it is owned, the same rate used for additional equipment costs is used for extended equipment costs. Idle equipment costs, however, are generally handled differently. The argument is that it costs less to own an idle piece of equipment than an operating piece of equipment. In contracts that address the topic, the rate for idled equipment and equipment on standby is usually discounted 50 percent. With regard to idle equipment, it is also important to remember that the contractor is usually thought to have an implied contractual obligation to mitigate the owner's damages. Thus, if the duration of a delay is known and is of such an extent that the cost of maintaining idled equipment on the site is greater than the cost of demobilizing the equipment and remobilizing the equipment when the delay is over, the contractor may have an obligation to do so. For example, the discovery of PCB-contaminated soil halts the excavation of foundations for a new office building. The owner informs the contractor that the shutdown will last 2 months while another contractor is brought onto the project to clean up the PCB-contaminated soils. The contractor was using a hydraulic excavator to perform excavation work. The rental cost of the excavator is $550 per day. The cost to demobilize and remobilize the equipment is $1200. In 2 months, rental charges will total almost $30,000, far more than the $1200 mobilization and demobilization cost. In this case, unless directed to do otherwise by the owner, the contractor should probably return the equipment to the rental company. If it does not, the owner could argue that the contractor did not take reasonable actions to mitigate the owner's costs, and could therefore deny reimbursement for the idled equipment.

Equipment cost escalation

There are a number of ways that a delay can result in an escalation of the contractor's equipment costs. As with labor costs, the source of the escalation may simply be that the delay postpones the performance of the work into a period when equipment costs are higher. This may occur because the rental cost of the equipment goes up or because less expensive equipment is no longer available and more expensive equipment must be used. For example, Fig. 8.4 shows a contractor's plan for constructing a fuel storage facility. The contractor planned to work through the spring, summer, and early fall months to prepare the project for the arrival of several large, prefabricated tanks. The project was shut down during August and September, however,

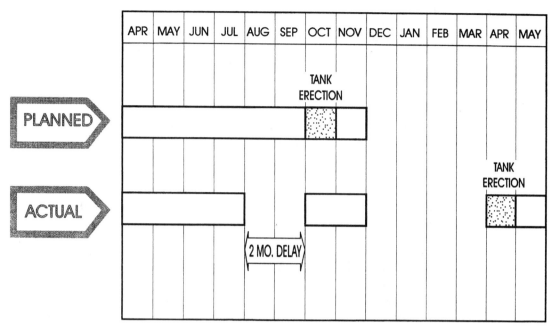

Figure 8.4 Contractor's plan for constructing a fuel storage facility.

because buried fuel tanks had been discovered on the site. The buried fuel tanks had leaked, contaminating soils in the vicinity of the new tanks. Work did not recommence until the beginning of October. Because of the delay, the contractor was not able to get the project ready for erection of the tanks before shutting the project down for the winter. Consequently, tank erection work was postponed until the following spring. Unfortunately, by the spring of the next year the crane the contractor had planned to rent to perform the tank erection work had been rented to a contractor on another project. The contractor was forced to rent a more expensive crane to prevent further delays to the project. The added rental cost was $250 per day. Thus, the 60-day delay attributed to the buried tanks resulted in the escalation of the crane cost. The cost effect on the contractor for the escalation alone was $250 per day for 30 days, for a total additional cost of $7500. Note in this case that the escalation is not paid for 60 days. Though the delay was 60 days in duration, the crane was only needed for 30 days. Thus, the escalated costs were only incurred for 30 days.

Delays or the discovery of contamination might affect equipment costs in other ways. For example, if it is discovered that, because of potential contamination, soils must be tested after they are excavated but before they can be transported offsite or used as fill material on site, it may not be possible to perform earth work as planned. The use of pans might be precluded. The work might have to be done using smaller pieces of equipment that do not outpace the owner's ability to test the materials before they are disposed of or placed.

Material Cost Effects

The three basic components of direct cost are labor, equipment, and material. Labor and equipment cost effects have been addressed. The last direct cost category is material. Material cost effects fall into three categories.

Additional material costs

When the discovery of contamination results in the need to procure additional materials, the cost of these added materials is recoverable. The cost effect is simply the amount invoiced by the manufacturer or supplier and paid by the contractor. Sometimes, suppliers offer discounts for prompt payment. Generally, the owner is entitled to benefit from a prompt-payment discount only if the contractor also takes advantage of the discount by making prompt payment. A few owners have added provisions to their contracts, however, indicating that they are entitled to the prompt-payment discount whether or not the contractor takes advantage of it.

Added materials may also be required as a result of a delay. For example, the extended use of battery-powered warning devices such as lights may require that additional batteries be purchased. Delaying work into the winter may require the purchase of blankets, hay, or other materials commonly used for cold-weather concrete pours.

Material cost escalation

If a compensable delay to a project occurs such that materials needed for the project are purchased for a higher price than expected, the contractor may be able to recover these added costs. A common example is concrete. Concrete cannot be purchased in advance; it must be purchased when it is to be used. Consequently, if a delay forces a contractor to perform concrete work later than planned and the work is performed when concrete costs are higher, the contractor is usually able to recover these added costs. The effect of material price increases is calculated in the same manner as labor cost escalation effects.

Material storage costs

Hand in glove with material cost escalation is the cost of material storage. One way to avoid material price increases is to purchase the material at the bid price and then store the material until it is needed. It is also sometimes necessary to take delivery of specially fabricated items and then store the items until they are actually needed. In either case, if inadequate storage exists onsite and the materials must be stored elsewhere due to a compensable delay, then the costs of acquiring this additional storage space are recoverable. The compensable duration of the storage is the same as the compensable delay that necessitated the storage. The cost effect is simply the number of days of storage times the daily rate. The number of days of storage is verified by schedule analysis. The rate is the rate invoiced by the storage facility and paid by the contractor.

Overhead and Other Markups

In addition to the direct cost effects of labor, equipment, and material, cost effects may include amounts for such costs as overhead, profit, bond, interest, and other costs. Whereas direct costs are often obvious, the cost effects associated with overhead or G&A (general and administrative), profit, etc., are less so. Consequently, the cost effects of these items are often a subject of debate. The cost of overhead, profit, bond, and interest is often expressed in terms of a percentage markup on the direct cost effects. It is not uncommon for these percentage markups to be specified in the contract. In fact, specifying these percentages can reduce the number of disagreements associated with these costs.

Overhead

Overhead costs are not easily attributable to a single item of work, but are costs incurred to support the management and administration of the project and the contractor's operations as a whole. For example, contractors often mobilize a trailer to the construction site to provide office space for the management of the project. The cost of this office cannot be attributed to any specific work activity, but is a cost attributable to the overall management and administration of the project. Particular overhead costs cannot be attributed to specific work activities, but there is little question that overhead costs are incurred and necessary for the proper execution of the work. The direct costs are usually marked up by some percentage to reflect the reasonable share of the contractor's overhead costs that must be borne by the added work.

Overhead comes in two basic varieties. The first is field overhead. *Field overhead costs* are those costs incurred on or near the project site that can be attributed specifically to the project, but not to any one work activity on the project. The construction trailer previously mentioned is a good example of a field overhead cost. The second type of overhead is home office overhead. *Home office overhead* represents the cost of managing and administering the contractor's operations as a whole. Home office overhead costs cannot usually be attributed to a single project. The contractor's president's salary is a good example of a typical home office overhead cost. Home office overhead costs also include the cost of services provided to various projects that can be handled more efficiently by the home office. For example, while payroll checks could be prepared by each project field office, payroll services are usually handled most efficiently by the company's home office and then distributed among a contractor's projects.

Since it is usually not possible to assign a particular overhead cost to a specific work activity, it is often difficult to identify the overhead costs that will increase as a result of the discovery of contamination on a project site. It is this difficulty in establishing a cause-and-effect relationship between the discovery of a problem and increased overhead costs that can make the calculation of overhead cost effects challenging and controversial. Complicating matters further, the method used to calculate overhead cost effects varies with the nature of the effect the discovery of contamination has on the performance of work on the project. The best way to address this variation is to look at each possibility separately.

The discovery of contamination results in added work, but no delay to the project.
The simplest situation occurs when the discovery of contamination adds work to the contract, but does not result in an overall project delay. For example, if, as a result of the discovery of contaminants in the soil, the owner asks the contractor to fence off a portion of the project site but requires no other action on the part of the contractor, then the work added to the contract is the added fencing. In this example the fencing work is not critical to the schedule, and the addition of a small amount of fencing will not result in a delay to the project. Clearly, the contractor is entitled to recover the added cost of the labor, equipment, and materials used to erect the additional fencing. Do the contractor's overhead costs increase as a result of this small change? The answer is perhaps debatable for this small change. Since typically most of a contractor's management and administrative personnel are salaried and are not paid hourly, and an additional person will probably not be hired to administer the change, it could be argued that there is no increase in overhead labor costs attributable to the change. Likewise, no additional office space or utilities are required. In sum, one could probably argue that, except for the cost of the paper and ink used to prepare the change-order documentation, no increase in the contractor's overhead costs may be associated with such a small change. Fortunately, most people in the construction industry don't waste a lot of time on these sorts of arguments. If looked at from a broader perspective, there is little doubt that as the volume of a contractor's work increases, management and administrative costs increase as well, though perhaps not proportionately or in lock-step. In fact, the connection between the volume of work and management and administrative costs is so obvious that many contracts specify that a contractor is entitled to some markup for overhead on the direct costs of a change. The magnitude of this specified overhead varies from a few percent to 15 or 20 percent, or higher. Usually the contract does not distinguish between field and home office overhead. Rather, a single percentage markup is established for both. Some contracts also include a markup for profit in the percentage markup for overhead.

Sometimes the contract does not provide for a specific markup for overhead. In such cases the overhead rate may be calculated or may be available from the contractor's financial records or bid documents. Many contractors apply a standard markup for overhead to their cost estimate to come up with a portion of their bid price for a project. This overhead rate can sometimes be documented from the bid documents. If bid documents are escrowed, this overhead rate can be verified by the bid documents. The overhead rate may also be calculated. The calculation requires the contractor to tabulate all of its overhead costs. Under some contracts, certain kinds of overhead costs are not recoverable. Usually, entertainment and some marketing costs fit into this category. Many public contracts refer to the Federal Acquisition Regulations concerning the allowability of overhead costs. The other number needed to calculate an overhead rate is a tabulation of costs incurred to perform the work. By dividing the overhead costs by the actual direct costs to perform the work, an overhead rate can be calculated.

The discovery of contamination results in the addition of work to the contract and a delay to the project's completion date. Using the fence example, how would the calculation of overhead change if the installation of the added fencing causes a 1-week critical delay to the project? Unlike the added work alone, delays to a project clearly result in increases to overhead costs, particularly field overhead costs. As a result of the project delay, the contractor will have a job-site trailer on the project longer. It will have a project manager or superintendent on site longer as well. In fact, all field overhead costs that are a function of time will increase as a result of the project delay. Calculating the field overhead cost effect resulting from a delay is not difficult. Simply sum the time-related field overhead expenses and divide by the number of days over which those costs were incurred. This will yield a daily rate. Then multiply this daily rate by the number of days of delay. This calculation will yield the field overhead costs associated with a particular delay.

The issue really is not how to calculate field overhead costs, but whether or not a contractor is entitled to recover both an overhead amount calculated as a percentage of the direct costs of the change and field overhead calculated on a daily basis. Some might argue that this is "double dipping," recovering the cost for the same item twice, particularly when the field overhead amount calculated on a daily basis is itself marked up for overhead on a percentage basis. Others argue, however, that there are overhead costs associated with the overall volume of the work as well as overhead costs associated with the time of performance of the work. Thus, if the addition of work results in a delay, the contractor should be entitled to recover overhead on the basis of both a percentage markup and a daily rate. This latter argument is further supported by the observation that if a percentage markup for overhead provides adequate compensation for the change alone, then clearly this percentage markup is not sufficient to cover the cost of the delay as well. Though this latter argument is the stronger and more equitable, some legal jurisdictions have ruled that a percentage markup on the direct costs of a change also constitutes adequate compensation for the cost effect of the associated delay (as long as the delay is not a suspension, a topic that is addressed in the next section). The key factor is the contract: If the contract is clear about what compensation a contractor is due for a delay, then there will often be little question and little need for arguing.

The discovery of contamination results in a suspension of the project work and a delay to the project. If the scenario used in this section is changed so that instead of simply adding fencing work, the project is suspended for a week until the owner determines how to deal with the contaminated soils, what is the associated overhead cost effect? Since no work was added, only suspended, the question of whether or not the contractor is entitled to recover both the overhead markup on the direct costs of the change and the recovery of field overhead on a daily basis essentially disappears. Clearly, the contractor is entitled to recover its field overhead costs on a daily basis, and there is no risk of double dipping. In situations such as this, however, it is common for a contractor to request compensation for unabsorbed or extended home office overhead costs as well. The contractor may also seek compensation for home office overhead costs when the delay is the result of added work and not a suspension. While it is easy to see how a delay increases field office overhead costs, it is more difficult to see how a delay on one project affects a contractor's home office overhead costs. In fact, a delay may very well have little or no effect on home office overhead costs, but instead affect reimbursement for these costs. In other words, the home office overhead cost effect associated with a delay is not the result of an increase in these costs; these costs should not change much whether a project is delayed or not. Rather, the issue is that a delay, particularly a delay resulting from a suspension of work, prevents a contractor from generating sufficient revenue to cover its home office overhead costs. To understand this better, consider how a contractor covers its home office overhead costs each month. Every month, a contractor incurs home office overhead costs for rent, salaries, utilities, and other, essentially fixed, costs. Every month a contractor also submits a bill to the owner of the project for the cost of the work performed that month. Once the contractor receives payment for this work, the money received is used first to cover the direct project costs of labor, equipment, material, and field management and administration. From the funds that remain, the contractor must cover the costs of its home office. In essence, home office overhead costs are recovered as a markup on the direct costs of performing the project work. Now consider what happens if work on a project is suspended. Because the contractor can do no work, there are no billings for the period of the suspension. Because there are no billings, there is no percentage markup on the direct costs of the project work available to cover the continuing home office overhead costs incurred during the period of the suspension. Since these costs neither disappear nor are reduced because of a project delay, the contractor must cover these costs some other way, either by borrowing from other projects or by dipping into its pockets or a line of credit. When a contractor seeks to recover the costs of home office overhead as a result of a delay, it is essentially seeking reimbursement for the home office overhead costs that were not covered by the suspended project during the period of the suspension.

The approach most often used to calculate unabsorbed home office overhead is to employ an apportionment formula of some kind. The most popular is a formula called the *Eichleay formula*. The Eichleay (pronounced eye-clay) formula comes from a court case involving the federal government and the Eichleay Corporation. Essentially, this case established the appropriateness of using the Eichleay formula to calculate unabsorbed home office overhead costs. As noted, the Eichleay formula is an apportionment formula. It apportions the contractor's overall home office overhead costs to a project based on the value of the project relative to the overall volume of the contractor's work. The formula is:

$$\frac{\text{Project value, including change orders}}{\text{Contractor's total revenue for contract period}} \times \text{home office overhead costs} = \text{allocable overhead}$$

$$\frac{\text{Allocable overhead}}{\text{Actual contract duration}} = \text{daily home office overhead rate}$$

Daily rate × days of delay = unabsorbed home office overhead cost effect

The Eichleay formula is not without its critics, and its use has not been uniformly supported by the courts. In situations where the project work was suspended, the suspension resulted in a critical project delay, and the contractor was not able to obtain other work to make up for the suspended work, at least on federal projects, the courts have been supportive of the use of the Eichleay formula. If the contractor's work was not suspended, however, and the delay resulted from extra work, or the contractor was able to substitute other work for the suspended work, or the contract specifically forbids the recovery of home office overhead costs for delay, the outcome is less clear.

Other formulas besides the Eichleay formula can be used to calculate a daily rate for home office overhead. One of these alternatives is known as the *Canadian method*. One problem with the Eichleay formula is that it is designed to be used in hindsight. If the parties are trying to negotiate a change before the work is actually performed and the resulting delay is incurred, then the Eichleay formula might not be appropriate. The Canadian method can be used in the same situation where an Eichleay formula might be used, but is particularly useful when negotiating changes before the changed work is actually performed.

Under the Canadian method, the focus is on the bid documents. This method is therefore facilitated by the escrow of bid documents. The cost component of the contractor's bid that relates to home office overhead is identified. Sometimes this number is entered under the heading of G&A. The estimated cost of home office overhead (making sure no field office costs are included) is then divided by the number of days of planned contract performance. This provides a daily rate. This daily rate is then multiplied by the number of days of delay to arrive at the cost of home office overhead. The Canadian method is not widely used. In situations where it is not possible to use the Eichleay formula or when some party objects to its use, the Canadian formula may provide an acceptable alternative.

The third alternative, and one way to address the criticism that the use of the Eichleay formula in situations where the delay is caused by the addition of work rather than a suspension results in double recovery, is to reduce the amount of the Eichleay damage by the amount of overhead paid on direct costs. For example, consider a critical project delay that resulted from the addition of work to the contract. This work included the removal of leaky fuel tanks discovered on the project site during excavation. The contractor's cost proposal prepared after the work was performed included an overhead markup of 15 percent on the direct costs of removing the tanks. This markup totaled $15,000. The contractor also included home office overhead costs calculated using the Eichleay formula totaling $23,000. If one argues that Eichleay costs are duplicative of the overhead markup, then one way to eliminate the duplication is to subtract the amount associated with the overhead markup from the calculated Eichleay costs. In fact, this approach has been used by some federal government agencies and others.

Whenever there is a potential for argument or controversy, as there is with regard to the recovery of home office overhead costs when the delay that gave rise to these costs is the result of added work, it is generally a good idea to address the controversial subject in the contract before it even arises. Some have addressed the recovery of home office overhead costs by simply writing contract provisions that forbid its recovery in any case. Such a contract clause is a modified form of a no-damage-for-delay clause. As a consequence, it increases the risk associated with the contract to potential bidders. This may result in higher prices. It is also unclear whether such a clause would always be enforceable. Potentially, the same kinds of arguments used to skirt a typical no-damage-for-delay clause might be used in an attempt to get around a clause that denied the recovery of home office overhead costs in all instances. The success of such arguments is hard to predict. A more palatable alternative might require that any Eichleay damages be offset by the overhead markup on the direct costs of the change. Such a clause might be perceived as being more fair and simply an attempt by the owner to reduce the potential for dispute.

Profit

In addition to overhead, a contractor is typically entitled to recover some amount for profit. Profit is generally calculated as a percentage markup on the total direct and indirect costs of the change. It generally ranges from 5 to 10 percent of these costs. Some contracts provide for variable profit rates based on the risk of the work and the magnitude of the change. Generally, the lower the risk and the larger the change, the lower will be the profit markup allowed.

Some contracts may also limit the recoverability of profit. For example, many suspension clauses prohibit the recovery of profit on cost effects attributable to the suspension.

Bond

Unless otherwise indicated in the contract, a contractor is generally allowed to mark up the final cost of a change for its bond premium. Bond premiums are charged as a percentage of the value of the contract value. A bond premium of around 1 percent is not uncommon.

Interest

The recoverability of interest is a function of the contract. For example, federal contracts allow the recovery of interest expenses only if a contractor is successful in recovering on a properly certified claim and then only from the date of certification. The question of the recoverability of interest is a legal one and depends on the jurisdiction in which the contract is interpreted, the contract, and other legal considerations. Both parties should clarify this area before the contract is executed.

Certification

This is to certify that this claim is made in good faith; that the supporting data are accurate and complete to the best of (the contractor's) knowledge and belief; and that the amount requested accurately reflects the contract adjustment for which (the contractor) believes the Government is liable.

By:_____

Date:___/___/___

Figure 8.5 Sample certification.

Certification

Owners are often concerned that cost proposals or claims submitted by contractors may be inflated. To address this concern, some owners have begun requiring contractors to certify the accuracy of the information included in the claim. An example of such a certification is provided in Fig. 8.5. If it is discovered that a certified claim is in fact inflated and includes costs that were not actually incurred, then the owner may assert that the claim is fraudulent. The submittal of a fraudulent claim could result in criminal prosecution of the contractor.

Chapter 9

Dealing with Disputes Arising from the Discovery of Contamination

The previous chapters have been devoted to the goal of preventing the kinds of problems that might lead to a dispute among the contracting parties related to the presence or discovery of contamination on the site. While the risk of a dispute can be reduced, it can never be eliminated completely. There will always be something to argue about on a construction project, particularly on a project where contamination is present. The objective of this chapter is to help assure that these inevitable disputes are resolved as quickly, efficiently, and amicably as possible.

The quickest, most efficient, and most amicable resolution will almost always occur when the dispute is resolved onsite during the performance of the work. Usually, the farther removed the forum for resolution of the dispute is from the project site and time of performance, the more difficult, costly, and contentious the resolution will be. Thus, the basic goal of any dispute resolution process is or should be to resolve disputes on the project site while the project is under way.

Dispute resolution procedures fall into two broad categories: nonbinding and binding.

Nonbinding Dispute Resolution Techniques

Nonbinding dispute resolution techniques are approaches to resolution that use methods where the parties do not enter into them knowing that this is the last chance. As the name clearly states, the process is "nonbinding." Consequently, these approaches are easier to agree to and more expeditious than a binding form of dispute resolution. There is less reluctance to attempt them, since the parties recognize that they have alternative avenues in case they are dissatisfied with the outcome of the process. For example, negotiation is a dispute resolution technique (in fact, probably one of the most commonly used and successful ones). It does not by design yield a result that is binding on the parties. Nobody makes a decision at the end of the session that everyone must live by. Instead, it is an opportunity for the parties to the dispute to sit down together in an attempt to talk through their differences in the hopes that a resolution acceptable to all will be reached. If the parties are able to find such a resolution, then the negotiation is successful, at least from the perspective of the resolution of the dispute. If not, then the dispute remains unresolved and the parties must fight on. Contrast this with binding arbitration or a trial. After the parties have had a chance to make their arguments, a judge, a jury, or a panel of arbitrators makes a decision that resolves the dispute and is binding on the parties.

The nonbinding forms of dispute resolution are sometimes grouped under the heading of ADR, which stands for Alternative Dispute Resolution. The general categories of nonbinding dispute resolution are negotiation, mediation, nonbinding arbitration, dispute review boards, minitrials, and the use of a settlement judge or court-appointed master. There are others, but these approaches are the most common.

Negotiation

Negotiation is the most common approach to the resolution of disputes on a construction project. Most disputes are resolved through simple negotiation. The parties to the dispute sit around a table at an agreed-upon time and place and hash out their differences. The first negotiation sessions should occur on the site among the onsite representatives of the parties to the dispute. If the onsite representatives cannot resolve the dispute through negotiation, then the dispute should be escalated to the next level. Escalation should continue until the dispute reaches the highest management levels of the parties involved. Many state departments of transportation have actually set forth this escalation process in their standard contract provisions. The contractor's superintendent and the state's resident engineer first try to resolve their differences. If they cannot, then, within a set period of time, the contractor must forward the dispute to the next level in the organization, often the district engineer. Within a certain amount of time, typically 30 to 90 days, the district engineer must respond and another attempt at a negotiated resolution is made. Beyond the district level, the dispute may be forwarded to the central office, the chief engineer, a "claims committee," or some other forum. At each step in the process, specific time limits are set and continued opportunities are provided to resolve the dispute by negotiation. Usually only as a last resort and after each step in the process has failed does a dispute at a state transportation agency using the procedures just described end up in the courts, before a binding arbitration panel, or before some form of contract appeals board.

The procedures used by state transportation agencies are illustrative of some of the techniques that can be used to improve the chances for a successful negotiation. A description of these techniques and others follows.

Time limits. There is a tendency to postpone the resolution of a dispute to the end of the project, once the work that gave rise to the dispute is completed. The old problem moves to the bottom of the "in box" to make way for new problems. It is retrieved only when the project is over and there are no new problems to consume the project manager's time. Unfortunately, by the end of the project, important details have been forgotten, crucial documents have been misplaced, and those most familiar with the dispute may have gone on to other projects. Recovering all this lost information may be impossible, but at a minimum is time consuming and costly. In addition, the lack of important information may make it more difficult for the parties to resolve their differences. For these reasons it is important to resolve disputes as they develop. In addition, unresolved disputes can poison relationships on a job, making it impossible to resolve later disputes that arise on the job. The sooner outstanding issues are resolved, the sooner the parties can put those issues behind them and focus on getting the project done on time, within budget or profitably, and in accordance with the plans and specifications.

Because there is a natural tendency to procrastinate on the resolution of disputes and because timely resolution often results in a better project, it is important to establish reasonable time limits that force the parties to address the dispute when it occurs and to provide as many opportunities as possible to resolve the dispute during the course of the project. Thus, as discussed in Chap. 3, a contract should contain a notice provision requiring the contractor to provide timely notice that a problem has occurred. This notice provision should also require that the contractor provide a detailed written description of the problem, including date of discovery, location information, appropriate contract references, and an analysis of the associated time and cost effects. Once provided to the owner or its representative, the contract should allow a reasonable period of time for evaluation. During this time, the owner and its advisors should review the contractor's request, inform the contractor immediately of any required information that is missing, gather information necessary to draft a response, and then prepare a written response to the contractor's notice. All parties involved in the dispute should enclose the documents that support their respective positions with their written responses.

If the problem is not resolved at this point, then the stage has been set for a truly effective negotiation session. Each side understands the other's position, the facts that support that position, and the strengths and weaknesses of both their position and those of the other parties to the dispute. All of this will have been accomplished while the project is ongoing and the issues that gave rise to the dispute are still fresh in everyone's mind. Also, it is generally true that by requiring the parties to the contract to address their disputes quickly, there may be more flexibility. By addressing problems expeditiously, opinions and positions have not had a chance to become fixed. This usually facilitates a quicker and more amicable resolution.

Preparation. Gamesmanship is no substitute for preparation. Oftentimes, the preparation for a negotiation becomes an exercise in structuring various games to be played by the negotiating team. Raising one's voice, pounding the table, setting up the straw man, and playing good-guy/bad-guy are all games that have been played and will continue to be played in an effort to intimidate or trick one party into agreeing to a position it might not otherwise accept. Ultimately, however, no amount of gamesmanship can mask a position that is poorly researched, poorly presented, and poorly argued, particularly when the other parties to the dispute have done their homework.

It is often helpful to think of a negotiation as a debate. Thus, it is important not only to prepare your own position, but to anticipate the other side's arguments and prepare well-reasoned and well-supported rebuttals. Also keep in mind that you are not simply presenting your position, but trying to convince the other parties to the dispute that your position is the right one. So don't hesitate to bring along pictures, charts, graphs, and other visual aids and documentation to support your position.

Stick to business. It is easy to personalize a dispute. For example, when a contractor's estimate to perform extra work seems high, too often the owner concludes not that the scope is poorly defined or that the work is especially risky or even that the contractor made a mistake, but that the contractor is an individual of questionable character who low-balled the job in the first place and is now trying to make up the difference through changes. Now, rather than settling with the contractor based on the merits of the contractor's position, the owner resists settling for fear that it will be caving in to some form of extortion by the contractor. The owner has personalized the dispute and, as a result, made settlement less likely and litigation more so.

Avoid the trap of personalizing the dispute by focusing on the facts. For example, if the contractor's estimated cost for a change seems high, ask for a copy of the contractor's estimate. Once received, review the estimate to see if it is accurate. If it is not, and you can prove it, you have a valid (and powerful) negotiating position. In fact, this is a much more powerful position than an argument based on a contractor's alleged criminality or whether the contractor low-balled its bid.

Break it up. Always keep in mind that every dispute potentially has three pieces. The first piece is entitlement. Was a change actually made for which the contractor is entitled to additional compensation and a time extension? The answer to this question is based on the contract, the law, and the facts of the dispute. If there is entitlement, then the next question is what were the impacts of the change. Was there a delay? Was there extra work? If there were effects, then the last question concerns the costs of the change.

Disputes on construction projects can often seem overwhelming given the number and technical nature of the issues that are encountered (particularly when site contamination is one of the issues), the complexity involved in measuring the effects of the problems encountered, and the large amounts of money that can be involved. The magnitude and complexity of the issues can make negotiation difficult, particularly when the parties are unprepared. One strategy for making negotiation of a dispute on a construction project more manageable is to break the discussions up into smaller, more manageable parts. For example, it is pointless to argue delays and costs until questions of entitlement have been resolved. Given this fact, the first negotiation session might be devoted to issues of entitlement. Once agreement is reached on entitlement, the discussion can shift to impacts, and then costs. Alternatively, if there are a number of issues in dispute, perhaps they can be broken down into groups and addressed three or four issues at a time. This approach can be helpful even if all issues are not ultimately resolved. It gives the parties an opportunity to narrow the field to just those issues that are truly in dispute and to resolve all others.

Keep in mind that if you break the dispute up into manageable pieces to be addressed in a number of sessions, it will be necessary to establish an agenda for each meeting and someone will have to keep minutes or at least summarize the results of each session. An agenda is a good idea for any negotiation meeting.

Trust but verify. There often comes a point in the negotiations when one side presents a fact that significantly bolsters its argument. The good news is that this revelation can become the basis for a settlement of the dispute. It is important, however, to verify this fact before an agreement is reached based on this fact. It is often true that a fact can be too good to be true, particularly if the dispute has been around for some time. There are few things more disheartening than compromising your position based on a fact presented by the other side that turns out to be unfounded or skewed in some way.

Provide plenty of opportunities to negotiate. Most disputes are resolved by negotiation. Don't put off the inevitable. Provide as many opportunities as possible as soon as possible to resolve the dispute through negotiation. If the dispute can be resolved by settling on the courtroom steps, it could have been resolved a lot sooner and a lot less expensively earlier.

In summary, odds are that your dispute will be settled by negotiation. Increase these odds by providing as many opportunities as possible to resolve the dispute by negotiation, by escalating negotiations to successive management levels until resolution is achieved, by setting realistic time limits between the time a contractor documents its position and the date the owner must respond, by splitting complicated issues into smaller parts, by sticking to facts and to business and not personalizing the dispute, by being prepared, and by using negotiating techniques to enhance your position, not as a substitute for a logical and persuasive position.

Mediation

Sometimes the parties by themselves are not able to reach a negotiated resolution to their dispute. This might be because the dispute is too old and the parties too polarized or because the issues are too complex or gray to allow the parties to resolve them on their own. An ever more popular approach to the resolution of such disputes is to enlist the services of an independent third party to help facilitate the negotiations. This facilitated negotiation is often called a mediation, and the facilitator is often called a mediator.

The approach taken to mediation varies with the mediator, but usually proceeds in the following manner. Once the parties agree to mediation, typically they will enter into a mediation agreement, select a mediator, and set up a time and place for the mediation. Shortly before the mediation, each side will provide a package of materials that sets forth its position on the dispute for the mediator. For large disputes, these packages can be quite voluminous. Typically they set forth in some detail the basis for each party's position on the questions of entitlement, impact, and cost. The mediator reviews these materials before the scheduled mediation date and on occasion may request supplemental information.

On the day of the mediation, the parties usually convene first for a joint session. At this joint session, each party to the negotiation is given an opportunity to present its position to the mediator and the other participants. Each participant's entourage typically includes someone capable of binding the party to any negotiated settlement, legal counsel, appropriate personnel from the project, and any experts or others assisting in analysis of the disputed issues. Depending on the size and nature of the dispute, presentations may take a few hours or all day. Following the presentations, the parties adjourn to separate rooms to meet alone with the mediator. During this time, the mediator will give each side an overview of how the mediator sees the dispute and also what the mediator believes are the chances for resolution. At this stage the mediator begins the "softening up" process. The mediator points out the weaknesses in your position and tells you to start thinking differently if you ever want to resolve the matter. The mediator advises you how he or she believes a judge or jury will rule on the issues. Once the mediator has set the tone for you to start thinking settlement, he or she meets with the other side and goes through the same process with them. The process continues with alternate meetings by each side with the mediator. Each meeting is an attempt to bring the sides closer together. Obviously, the mediator has the common knowledge of the real positions of both sides. This allows the manipulation of the parties toward a common ground. It is not uncommon for a mediator to take a position with which you cannot agree. He or she may do this to create the belief that a judge or other impartial party may also see the dispute in the same light. This may give rise to doubts about the strength of your position and motivate you toward a settlement position.

When dealing with a mediator, it is important to keep at least one thing foremost in your mind. A mediator is not a judge. Consequently, a mediator is not looking for justice or equity. A mediator is primarily, if not exclusively, interested in achieving settlement. Other than the fact that you also desire to settle the dispute, the mediator does not represent your interests.

The keys to being successful in mediation are similar to those described for successful negotiation. Primarily, however, it is important to be prepared, to know clearly the strengths and weaknesses of your position, and to stick to the facts.

Nonbinding arbitration, minitrials, settlement judges, and court-appointed masters

Typically as an alternative to mediation rather than in addition to it, the parties may elect to use (or be persuaded by a judge to use) some form of an alternative dispute resolution process in which an independent third party actually renders an opinion or recommendation regarding the merits of each party's position in a dispute. Though it is nonbinding, such an evaluation may carry sufficient weight to induce one or all of the parties to modify their positions and facilitate a negotiated resolution. This process can take many forms and is known by many names. If it is employed during the course of the project, a panel known as a disputes review board may be convened by contract to review and render decisions on disputes as they arise. The parties may elect to try a minitrial with a "judge" or neutral in order to avoid the cost of the full litigation process. A court may appoint a retired judge or another judge not involved with the case to help the parties move toward settlement. The neutral, settlement judge or court-appointed master will typically assume the same role as the disputes review board, reviewing the positions of the parties and rendering a nonbinding judgment concerning the merits of each side's position. Again, the objective is to persuade the parties to move off their entrenched positions and make settlement possible.

Dispute review boards (DRBs) have been used successfully on many projects to facilitate the resolution of disputes quickly and amicably. The first step in the implementation of a dispute review board is to place a provision in the contract that requires their use and provides for funding. See Chap. 4 for an example of such a clause. Once the contractor has been selected, the next step is to choose the members of the panel. Typically, one member of the three-member panel is chosen by the contractor, one by the owner, and one by the two members selected by the owner and the contractor. Many different varieties of DRBs are presently in use. While not all permutations can be discussed, the elements which appear to be most successful will be highlighted.

The panel may be required to visit the project site on a periodic basis to become familiar with and to monitor the progress of the work. Either on an as-needed basis or at selected intervals corresponding with site visits, the disputes review board is convened to review disputes elevated to that level by the parties to the contract. Either the owner or the contractor can bring an issue before the DRB at any time. Normally this is done by making a request to the chairman of the DRB. When the DRB convenes to hear a dispute, each side is given an opportunity to present its position to the board. All necessary documentation and analysis is also provided by each side to the board. In the most efficient boards, the positions and all supporting documentation are submitted to all board members at least 10 working days in advance of the hearing.

During the hearing, the board may ask questions and may request additional information. Within a short time of convening to review a dispute, the board provides the results of its review in writing to the parties, with appropriate reasoning to support the conclusions reached. DRB decisions are normally referred to as "recommendations," which means exactly what it says—the process is nonbinding. Though the decision of the DRB is nonbinding, in many instances, the board's decision is admissible as evidence in court. The objective of a DRB is to provide each party to the dispute with a knowledgeable and impartial evaluation of its position.

One reason that DRBs have been effective is that the board members usually have considerable experience in construction and do not have to be educated concerning the nature of the issues. Also, by meeting throughout the life of the project, board members develop a rapport with the parties and establish a degree of mutual respect. In addition, because the board visits the site and understands the work, it is already familiar with the issues that may be contested.

Minitrials are a condensed version of the "real thing," an actual trial. In a minitrial a "judge" or neutral is selected by both parties. Prior to the minitrial the parties submit their positions, with all supporting documentation, to the neutral. The format of the minitrial is established by the parties with the advice of the neutral. Usually, both sides present their positions in a testimonial form, but without the swearing of oaths. Often, cross-examination is not allowed in order to expedite the process. After each side has made its presentation, some brief rebuttal is allowed. At that point the "trial portion" of the process is adjourned. The neutral takes a day or two to formulate an opinion and then meets with principals from both sides to discuss the neutral's findings. At this stage the neutral begins to act like a mediator by trying to help both parties find a common ground based on the neutral's assessment of the facts of the case and the applicable law. Unlike a mediator, who is interested primarily in a resolution, the neutral is attempting to find a common ground based on the equity of the situation. Normally, information used at a minitrial is inadmissible in any subsequent arbitration or litigation.

Settlement judges and court-appointed masters work somewhat differently than dispute review boards. Their use is not usually established by contract or established before work begins. Their use is usually by mutual agreement of the parties to the dispute after negotiations to resolve the dispute have failed. They may be provided by the court or obtained separately by the parties. However established, the procedure by which they accomplish their work is similar. The judge is appointed and a date and time set for a "hearing," sometimes referred to as an ADR or minitrial. Prior to the hearing, each party to the dispute submits a detailed package setting forth its position on the issues in dispute. At the ADR itself, each side is given an opportunity to present its arguments. Typically, no more than one to two days are needed to complete these presentations. At the conclusion of the presentations, the "judge" renders an opinion. This opinion is not binding and usually is not admissible in court.

Nonbinding arbitration is essentially identical to a dispute review board, but is not usually established or mandated by the contract. It is usually conducted by mutual agreement of the parties after negotiations have failed. The panel's decision carries as much weight in court as the parties agree.

The great strengths of nonbinding approaches to the resolution of disputes are the relative speed with which they can be accomplished, the relatively low cost of using them, and the high success rate. The chief weakness is that success is not assured. Because the results are nonbinding, the possibility always exists that nonbinding dispute resolution techniques will be unsuccessful. In addition, their use can actually be harmful if one of the parties is intentionally using these techniques to delay resolution of the dispute.

Binding Arbitration, Courts, and Contract Appeals Boards

If the parties are not able to resolve their dispute through negotiation or some other nonbinding dispute resolution procedure, the dispute must ultimately go before a judge, a jury, an arbitration panel, or a board of contract appeals. The strength of these forums is the finality and binding nature of the decisions reached. In addition, the discovery process and legal precedent may actually facilitate the settlement of disputes. The weaknesses of these forums include the length of time required to reach a decision, their high cost, and the animosity that develops between the parties as a result of the proceedings. The differences between arbitration and the courts are many, though in some cases the practical benefits of arbitration over the courts may be limited. The use of binding arbitration may be mandated by contract or agreed to by the parties. Typically, though not necessarily, it is provided through the American Arbitration Association (AAA). Once one or both of the parties to the dispute concludes that further negotiations are fruitless, the dispute is submitted for arbitration. A panel is selected. Panel selection can be accomplished several ways, but typically selections are made from a list of arbitrators provided by the AAA. Panels may have one member for smaller cases and three for larger cases. On panels with three members, the lead arbitrator is often an attorney, but there is no requirement that this be the case. It is perfectly acceptable for the arbitration panel to have no attorneys as members.

The conduct of an arbitration proceeding is usually very similar to that of a trial, though often much less formal. There is usually a court reporter. Discovery may be allowed. Each side presents fact and expert witnesses. Cross-examination is permitted. There are usually few limits on the evidence and testimony that is admitted. Once each side has presented its case, the arbitrators are usually allowed 30 days to reach a decision. Unless otherwise established, the decision is usually short, with little, if any, explanation.

For small disputes that can be presented in one or two days of hearings, arbitration can have real benefits over litigation in the courts. A panel can be convened relatively quickly and hearings scheduled expeditiously. Because the proceedings are less formal and typically are conducted without the cumbersome discovery process associated with litigation, presentation of the case can proceed more quickly. It is also believed that because the arbitrators are experienced and drawn from the construction industry, their decisions will be more informed and fair. Though AAA is in the midst of implementing reformed procedures, right now the arbitration of large disputes can end up being as time consuming and expensive as litigation. The reason for this is related to the method of selection and the types of individuals selected to sit on arbitration panels. First, in an industry as small as the construction industry in any particular region, it can be difficult to assemble an arbitration panel that is acceptable to all parties. In addition, because arbitration panelists are very often actively employed at jobs in the industry, the panel cannot meet continuously. Instead, dates must be fitted into the schedules of all panelists, counsel representing each side in the dispute, and the various witnesses. Sometimes this limits hearings to once a month at best. As a result, it is not uncommon for arbitrations conducted in this manner to stretch out for a year or more. Keeping attorneys and witnesses prepared for such an extensive period of time is quite expensive. If the panel also allows some form of discovery prior to scheduling hearings, arbitration can end up taking longer and costing as much as litigation, muting if not eliminating some of the advantages of arbitration.

Some also feel that the fact that arbitrators are not usually required to explain their decisions is a significant drawback. No lessons are learned, and the parties are provided with only limited guidance for future conduct. On the plus side, it is very difficult to overturn an arbitration panel's decision, in that the rights to appeal are extremely limited. This gives such a decision much more finality.

The list of weaknesses in the country's legal process (including contract appeals boards) is extensive. It is expensive. It is time consuming. It is contentious. It is stressful. And, particularly on very technical cases, the decisions rendered by a jury or judge may not be entirely satisfactory (hence the development of arbitration). Also, given the extensive right of appeal, decisions rendered by lower courts may be disputed for years. Government contract appeals boards may take years to render a decision.

The litigation process does, however, possess some desirable qualities. The discovery process is one example. Though they are time consuming and expensive, document productions, interrogatories, and depositions help get all the facts on the table before the trial actually begins. Once each side understands the strengths and weaknesses of each side and the legal arguments to be made, the chances for settlement should be enhanced. Indeed, it is relatively common for a case to be settled on the courthouse steps. In addition, because a court's rulings tend to follow the rulings of courts that preceded them, the decisions a court will make are somewhat predictable. This body of prior decisions, known as precedent, also helps to facilitate the settlement of disputes, since it can inform the parties in advance how their dispute may be looked upon by the courts.

Appendix

Requirements for Owners and Operators of Underground Storage Tanks*

Part 280—Technical Standards and Corrective Action. Requirements for Owners and Operators of Underground Storage Tanks (USTs)

1. The authority citation for part 280 is revised to read as follows:

   Authority: 42 U.S.C. 6912, 6991, 6991a, 6991b, 6991c, 6991d, 6991e, 6991f, 6991g, 6991h.

2. Part 280 is amended by adding subpart I consisting of §§280.200 through 280.230 to read as follows:

Subpart I—Lender Liability

§280.200 Definitions.

§280.210 Participation in management.

§280.220 Ownership of an underground storage tank or underground storage tank system or facility or property on which an underground storage tank or underground storage tank system is located.

§280.230 Operating an underground storage tank or underground storage tank system.

---

*Federal Register/Vol. 60, No. 173 /Thursday, September 7, 1995 /Rules and Regulations 46711.

§280.200 Definitions

(a) *UST Technicals,* as used in this subpart, refers to the UST preventative and operating requirements under 40 CFR part 280, subparts B, C, D, G, and §280.50 of subpart E.

(b) *Petroleum production, refining, and marketing.*

    (1) *Petroleum production* means the production of crude oil or other forms of petroleum (as defined in §280.12) as well as the production of petroleum products from purchased materials.

    (2) *Petroleum refining* means the cracking, distillation, separation, conversion, upgrading, and finishing of refined petroleum or petroleum products.

    (3) *Petroleum marketing* means the distribution, transfer, or sale of petroleum or petroleum products for wholesale or retail purposes.

(c) *Indicia of ownership* means evidence of a secured interest, evidence of an interest in a security interest, or evidence of an interest in real or personal property securing a loan or other obligation, including any legal or equitable title or deed to real or personal property acquired through or incident to foreclosure. Evidence of such interests include, but are not limited to, mortgages, deeds of trust, liens, surety bonds and guarantees of obligations, title held pursuant to a lease financing transaction in which the lessor does not select initially the leased property (hereinafter "lease financing transaction"), and legal or equitable title obtained pursuant to foreclosure. Evidence of such interests also includes assignments, pledges, or other rights to or other forms of encumbrance against property that are held primarily to protect a security interest. A person is not required to hold title or a security interest in order to maintain indicia of ownership.

(d) A *holder* is a person who, upon the effective date of this regulation or in the future, maintains indicia of ownership (as defined in §280.200© primarily to protect a security interest (as defined in §280.200(f)(I)) in a petroleum UST or UST system or facility or property on which a petroleum UST or UST system is located. A holder includes the initial holder (such as a loan originator); any subsequent holder (such as a successor-in-interest or subsequent purchaser of the security interest on the secondary market); a guarantor of an obligation, surety, or any other person who holds ownership indicia primarily to protect a security interest; or a receiver or other person who acts on behalf or for the benefit of a holder.

(e) A *borrower, debtor,* or *obligor* is a person whose UST or UST system or facility or property on which the UST or UST system is located is encumbered by a security interest. These terms may be used interchangeably.

    (1) *Security interest* means an interest in a petroleum UST or UST system or in the facility or property on which a petroleum UST or UST system is located, created or established for the purpose of securing a loan or other obligation. Security interests include but are not limited to mortgages, deeds of trusts, liens, and title pursuant to lease financ-

ing transactions. Security interests may also arise from transactions such as sale and leasebacks, conditional sales, installment sales, trust receipt transactions, certain assignments, factoring agreements, accounts receivable financing arrangements, and consignments, if the transaction creates or establishes an interest in an UST or UST system or in the facility or property on which the UST or UST system is located, for the purpose of securing a loan or other obligation.

(2) *Primarily to protect a security interest,* as used in this subpart, does not include indicia of ownership held primarily for investment purposes, nor ownership indicia held primarily for purposes other than as protection for a security interest. A holder may have other, secondary reasons for maintaining indicia of ownership, but the primary reason why any ownership indicia are held must be as protection for a security interest.

(f) *Operation* means, for purposes of this subpart, the use, storage, filling, or dispensing of petroleum contained in an UST or UST system.

§280.210 Participation in management

The term "participating in the management of an UST or UST system" means that, subsequent to the effective date of this subpart, December 6, 1995, the holder is engaging in decision making control of, or activities related to, operation of the UST or UST system, as defined herein.

(a) Actions that are participation in management.
  (1) Participation in the management of an UST or UST system means, for purposes of this subpart, actual participation by the holder in the management or control of decision making related to the operation of an UST or UST system. Participation in management does not include the mere capacity or ability to influence or the unexercised right to control UST or UST system operations. A holder is participating in the management of the UST or UST system only if the holder either:
    (i) Exercises decision making control over the operational (as opposed to financial or administrative) aspects of the UST or UST system, such that the holder has undertaken responsibility for all or substantially all of the management of the UST or UST system; or
    (ii) Exercises control at a level comparable to that of a manager of the borrower's enterprise, such that the holder has assumed or manifested responsibility for the overall management of the enterprise encompassing the day-to-day decision making of the enterprise with respect to all, or substantially all, of the operational (as opposed to financial or administrative) aspects of the enterprise.
  (2) Operational aspects of the enterprise relate to the use, storage, filling, or dispensing of petroleum contained in an UST or UST system, and include functions such as that of a facility or plant manager, operations manager, chief operating officer, or chief executive officer.

Financial or administrative aspects include functions such as that of a credit manager, accounts payable/receivable manager, personnel manager, controller, chief financial officer, or similar functions. Operational aspects of the enterprise do not include the financial or administrative aspects of the enterprise or actions associated with environmental compliance, or actions undertaken voluntarily to protect the environment in accordance with applicable requirements in 40 CFR part 280 or applicable state requirements in those states that have been delegated authority by EPA to administer the UST program pursuant to 42 UST 6991c and 40 CFR part 281.

(b) Actions that are not participation in management pre-foreclosure.

 (1) Actions at the inception of the loan or other transaction. No act or omission prior to the time that indicia of ownership are held primarily to protect a security interest constitutes evidence of participation in management within the meaning of this subpart. A prospective holder who undertakes or requires an environmental investigation (which could include a site assessment, inspection, and/or audit) of the UST or UST system or facility or property on which the UST or UST system is located (in which indicia of ownership are to be held), or requires a prospective borrower to clean up contamination from the UST or UST system or to comply or come into compliance with (whether prior or subsequent to the time that indicia of ownership are held primarily to protect a security interest) any applicable law or regulation, is not by such action considered to be participating in the management of the UST or UST system or facility or property on which the UST or UST system is located.

 (2) Loan policing and work out. Actions that are consistent with holding ownership indicia primarily to protect a security interest do not constitute participation in management for purposes of this subpart. The authority for the holder to take such actions may, but need not, be contained in contractual or other documents specifying requirements for financial, environmental, and other warranties, covenants, conditions, representations or promises from the borrower. Loan policing and work out activities cover and include all such activities up to foreclosure, exclusive of any activities that constitute participation in management.

  (i) Policing the security interest or loan.

   (A) A holder who engages in policing activities prior to foreclosure will remain within the exemption provided that the holder does not together with other actions participate in the management of the UST or UST system as provided in 280.210(a). Such policing actions include, but are not limited to, requiring the borrower to clean up contamination from the UST or UST system during the term of the security interest; requiring the borrower to comply or come into compliance with applicable federal, state, and local environmental and other laws, rules, and regulations during the term of the security interest; securing or exercising authority to monitor or inspect the UST or UST system or facility or property on which the UST or UST system is located (including on-site inspections) in which indicia of ownership are maintained, or the borrower's business or financial condition during the term of the security interest; or taking other actions to adequately police the loan or security interest (such as requiring a borrower to comply with any warranties, covenants, conditions, representations, or promises from the borrower).

(B) Policing activities also include undertaking by the holder of UST environmental compliance actions and voluntary environmental actions taken in compliance with 40 CFR part 280, provided that the holder does not otherwise participate in the management or daily operation of the UST or UST system as provided in §280.210(a) and §280.230. Such allowable actions include, but are not limited to, release detection and release reporting, release response and corrective action, temporary or permanent closure of an UST or UST system, UST upgrading or replacement, and maintenance of corrosion protection. A holder who undertakes these actions must do so in compliance with the applicable requirements in 40 CFR part 280 or applicable state requirements in those states that have been delegated authority by EPA to administer the UST program pursuant to 42 U.S.C. 6991c and 40 CFR part 281. A holder may directly oversee these environmental compliance actions and voluntary environmental actions, and directly hire contractors to perform the work, and is not by such action considered to be participating in the management of the UST or UST system.

(ii) Loan work out. A holder who engages in work out activities prior to foreclosure will remain within the exemption provided that the holder does not together with other actions participate in the management of the UST or UST system as provided in §280.210(a). For purposes of this rule, "work out" refers to those actions by which a holder, at any time prior to foreclosure, seeks to prevent, cure, or mitigate a default by the borrower or obligor; or to preserve, or prevent the diminution of, the value of the security. Work out activities include, but are not limited to, restructuring or renegotiating the terms of the security interest; requiring payment of additional rent or interest; exercising forbearance; requiring or exercising rights pursuant to an assignment of accounts or other amounts owing to an obligor; requiring or exercising rights pursuant to an escrow agreement pertaining to amounts owing to an obligor; providing specific or general financial or other advice, suggestions, counseling, or guidance; and exercising any right or remedy the holder is entitled to by

law or under any warranties, covenants, conditions, representations, or promises from the borrower.
   (c) Foreclosure on an UST or UST system or facility or property on which an UST or UST system is located, and participation in management activities post-foreclosure.
      (1) Foreclosure.
         (i) Indicia of ownership that are held primarily to protect a security interest include legal or equitable title or deed to real or personal property acquired through or incident to foreclosure. For purposes of this subpart, the term "foreclosure" means that legal, marketable or equitable title or deed has been issued, approved, and recorded, and that the holder has obtained access to the UST, UST system, UST facility, and property on which the UST or UST system is located, provided that the holder acted diligently to acquire marketable title or deed and to gain access to the UST, UST system, UST facility, and property on which the UST or UST system is located. The indicia of ownership held after foreclosure continued to be maintained primarily as protection for a security interest provided that the holder undertakes to sell, release an UST or UST system or facility or property on which the UST or UST system is located, held pursuant to a lease financing transaction (whether by a new lease financing transaction or substitution of the lessee), or otherwise divest itself of the UST or UST system or facility or property on which the UST or UST system is located, in a reasonably expeditious manner, using whatever commercially reasonable means are relevant or appropriate with respect to the UST or UST system or facility or property on which the UST or UST system is located, taking all facts and circumstances into consideration, and provided that the holder does not participate in management (as defined in §280.210(a)) prior to or after foreclosure.
         (ii) For purposes of establishing that a holder is seeking to sell, release pursuant to a lease financing transaction (whether by a new lease financing transaction or substitution of the lessee), or divest in a reasonably expeditious manner an UST or UST system or facility or property on which the UST or UST system is located, the holder may use whatever commercially reasonable means as are relevant or appropriate with respect to the UST or UST system or facility or property on which the UST or UST system is located, or may employ the means specified in §280.210(c)(2). A holder that outbids, rejects, or fails to act upon a written *bona fide,* firm offer of fair consideration for the UST or UST system or facility or property on which the UST or UST system is located, as provided in §280.210(c)(2), is not considered to hold indicia of ownership primarily to protect a security interest.

(2) Holding foreclosed property for disposition and liquidation. A holder, who does not participate in management prior to or after foreclosure, may sell, re-lease, pursuant to a lease financing transaction (whether by a new lease financing transaction or substitution of the lessee), an UST or UST system or facility or property on which the UST or UST system is located, liquidate, wind up operations, and take measures, prior to sale or other disposition, to preserve, protect, or prepare the secured UST or UST system or facility or property on which the UST or UST system is located. A holder may also arrange for an existing or new operator to continue or initiate operation of the UST or UST system. The holder may conduct these activities without voiding the security interest exemption, subject to the requirements of this subpart.

   (i) A holder establishes that the ownership indicia maintained after foreclosure continue to be held primarily to protect a security interest, within 12 months following foreclosure, by listing the UST or UST system or the facility or property on which the UST or UST system is located, with a broker, dealer, or agent who deals with the type of property in question, or by advertising the UST or UST system or facility or property on which the UST or UST system is located, as being for sale or disposition on at least a monthly basis in either a real estate publication or a trade or other publication suitable for the UST or UST system or facility or property on which the UST or UST system is located, or a newspaper of general circulation (defined as one with a circulation over 10,000, or one suitable under any applicable federal, state, or local rules of court for publication required by court order or rules of civil procedure) covering the location of the UST or UST system or facility or property on which the UST or UST system is located. For purposes of this provision, the 12-month period begins to run from December 6, 1995 or from the date that the marketable title or deed has been issued, approved and recorded, and the holder has obtained access to the UST, UST system, UST facility and property on which the UST or UST system is located, whichever is later, provided that the holder acted diligently to acquire marketable title or deed and to obtain access to the UST, UST system, UST facility and property on which the UST or UST system is located. If the holder fails to act diligently to acquire marketable title or deed or to gain access to the UST or UST system, the 12-month period begins to run from December 6, 1995 or from the date on which the holder first acquires either title to or possession of the secured UST or UST system, or facility or property on which the UST or UST system is located, whichever is later.

   (ii) A holder that outbids, rejects, or fails to act upon an offer of fair consideration for the UST or UST system or the facility or property on which the UST or UST system is located, establishes by such outbidding, rejection, or failure to act, that the ownership indicia in the secured UST or UST system or facility or property on which the UST or UST system is located are not held primarily to protect the security interest, unless the holder is required, in order to avoid liability under federal or state law, to make a higher bid, to obtain a higher offer, or to seek or obtain an offer in a different manner.

(A) Fair consideration, in the case of a holder maintaining indicia of ownership primarily to protect a senior security interest in the UST or UST system or facility or property on which the UST or UST system is located, is the value of the security interest as defined in this section. The value of the security interest includes all debt and costs incurred by the security interest holder, and is calculated as an amount equal to or in excess of the sum of the outstanding principal (or comparable amount in the case of a lease that constitutes a security interest) owed to the holder immediately preceding the acquisition of full title (or possession in the case of a lease financing transaction) pursuant to foreclosure, plus any unpaid interest, rent, or penalties (whether arising before or after foreclosure). The value of the security interest also includes all reasonable and necessary costs, fees, or other charges incurred by the holder incident to work out, foreclosure, retention, preserving, protecting, and preparing, prior to sale, the UST or UST system or facility or property on which the UST or UST system is located, re-lease, pursuant to a lease financing transaction (whether by a new lease financing transaction or substitution of the lessee), of an UST or UST system or facility or property on which the UST or UST system is located, or other disposition. The value of the security interest also includes environmental investigation costs (which could include a site assessment, inspection, and/or audit of the UST or UST system or facility or property on which the UST or UST system is located), and corrective action costs incurred under §§280.51 through 280.67 or any other costs incurred as a result of reasonable efforts to comply with any other applicable federal, state or local law or regulation; less any amounts received by the holder in connection with any partial disposition of the property and any amounts paid by the borrower (if not already applied to the borrower's obligations) subsequent to the acquisition of full title (or possession in the case of a lease financing transaction) pursuant to foreclosure. In the case of a holder maintaining indicia of ownership primarily to protect a junior security interest, fair consideration is the value of all outstanding higher priority security interest plus the value of the security interest held by the junior holder, each calculated as set forth in this paragraph.

(B) Outbids, rejects, or fails to act upon an offer of fair consideration means that the holder outbids, rejects, or fails to act upon within 90 days of receipt, a written, *bona fide,* firm offer of fair consideration for the UST or UST system or facility or property on which the UST or UST system is located received at any time after six months following foreclosure, as defined in §280.210(c). A "written, *bona fide,* firm offer" means a legally enforceable, commercially reasonable, cash offer solely for the foreclosed UST or UST system or facility or property on which the UST or UST system is located, including all material terms of the transaction, from a ready, willing, and able purchaser who demonstrates to the holder's satisfaction the ability to perform. For purposes of this provision, the six-month period begins to run from December 6, 1995 or from the date that marketable title or deed has been issued, approved and recorded to the holder, and the holder has obtained access to the UST, UST system, UST facility and property on which the UST or UST system is located, whichever is later, provided that the holder was acting diligently to acquire marketable title or deed and to obtain access to the UST or UST system, UST facility and property on which the UST or UST system is located. If the holder fails to act diligently to acquire marketable title or deed or to gain access to the UST or UST system, the six-month period begins to run from December 6, 1995 or from the date on which the holder first acquires either title to or possession of the secured UST or UST system, or facility or property on which the UST or UST system is located, whichever is later.

(3) Actions that are not participation in management post-foreclosure. A holder is not considered to be participating in the management of an UST or UST system or facility or property on which the UST or UST system is located when undertaking actions under 40 CFR part 280, provided that the holder does not otherwise participate in the management or daily operation of the UST or UST system as provided in §280.210(a) and §280.230. Such allowable actions include, but are not limited to, release detection and release reporting, release response and corrective action, temporary or permanent closure of an UST or UST system, UST upgrading or replacement, and maintenance of corrosion protection. A holder who undertakes these actions must do so in compliance with the applicable requirements in 40 CFR part 280 or applicable state requirements in those states that have been delegated authority by EPA to administer the UST program pursuant to 42 U.S.C. 6991c and 40 CFR part 281. A holder may directly

oversee these environmental compliance actions and voluntary environmental actions, and directly hire contractors to perform the work; and is not by such action considered to be participating in the management of the UST or UST system.

§280.220 Ownership of an underground storage tank or underground storage tank system or facility or property on which an underground storage tank or underground storage tank system is located

Ownership of an UST or UST system or facility or property on which an UST or UST system is located. A holder is not an "owner" of a petroleum UST or UST system or facility or property on which a petroleum UST or UST system is located for purposes of compliance with the UST technical standards as defined in §280.200(a), the UST corrective action requirements under §§280.51 through 280.67, and the UST financial responsibility requirements under §§280.90 through 280.111, provided the person:

(a) Does not participate in the management of the UST or UST system as defined in §280.210; and

(b) Does not engage in petroleum production, refining, and marketing as defined in §280.200(b).

§280.230 Operating an underground storage tank or underground storage tank system

(a) Operating an UST or UST system prior to foreclosure. A holder, prior to foreclosure, as defined in §280.210(c), is not an "operator" of a petroleum UST or UST system for purposes of compliance with the UST technical standards as defined in §280.200(a), the UST corrective action requirements under §§280.51 through 280.67, and the UST financial responsibility requirements under §§280.90 through 280.111, provided that, after December 6, 1995, the holder is not in control of or does not have responsibility for the daily operation of the UST or UST system.

(b) Operating an UST or UST system after foreclosure. The following provisions apply to a holder who, through foreclosure, as defined in §280.21(c), acquires a petroleum UST or UST system or facility or property on which a petroleum UST or UST system is located.

(1) A holder is not an "operator" of a petroleum UST or UST system for purposes of compliance with 40 CFR part 280 if there is an operator, other than the holder, who is in control of or has responsibility for the daily operation of the UST or UST system, and who can be held responsible for compliance with applicable requirements of 40 CFR part 280 or applicable state requirements in those states that have been delegated authority by EPA to administer the UST program pursuant to 42 U.S.C. 6991c and 40 CFR part 281.

(2) If another operator does not exist, as provided for under paragraph (b)(I) of this section, a holder is not an "operator" of the UST or UST

system, for purposes of compliance with the UST technical standards as defined in §280.200(a), the UST corrective action requirements under §§280.51 through 280.67, and the UST financial responsibility requirements under §§280.90 through 280.111, provided that the holder:
   (i) Empties all of its known USTs and UST system within 60 calendar days after foreclosure or within 60 calendar days after December 6, 1995, whichever is later, or another reasonable time period specified by the implementing agency, so that no more than 2.5 centimeters (one inch) of residue, or 0.3 percent by weight of the total capacity of the UST system, remains in the system; leaves vent lines open and functioning, and caps and secures all other lines, pumps, manways, and ancillary equipment; and
   (ii) Empties those USTs and UST systems that are discovered after foreclosure within 60 calendar days after discovery or within 60 calendar days after December 6, 1995, whichever is later, or another reasonable time period specified by the implementing agency, so that no more than 2.5 centimeters (one inch) of residue, or 0.3 percent by weight of the total capacity of the UST system, remains in the system; leaves vent lines open and functioning; and caps and secures all other lines, pumps, manways, and ancillary equipment.
(3) If another operator does not exist, as provided for under paragraph (b)(1) of this section, in addition to satisfying the conditions under paragraph (b)(2) of this section, the holder must either:
   (i) Permanently close the UST or UST system in accordance with §§280.71 through 280.74, except §280.72(b); or
   (ii) Temporarily close the UST or UST system in accordance with the following applicable provisions of §280.70;
      (A) Continue operation and maintenance of corrosion protection in accordance with §280.31;
      (B) Report suspected releases to the implementing agency; and
      (C) Conduct a site assessment in accordance with §280.72(a) if the UST system is temporarily closed for more than 12 months and the UST system does not meet either the performance standards in §280.20 for new UST systems or the upgrading requirements in §280.21, except that the spill and overfill equipment requirements do not have to be met. The holder must report any suspected releases to the implementing agency. For purposes of this provision, the 12-month period begins to run from December 6, 1995 or from the date on which the UST system is emptied and secured under paragraph (b)(2) of this section, whichever is later.
(4) The UST system can remain in temporary closure until a subsequent purchaser has acquired marketable title to the UST or UST system or facility or property on which the UST or UST system is located.

Once a subsequent purchaser acquires marketable title to the UST or UST system or facility or property on which the UST or UST system is located, the purchaser must decide whether to operate or close the UST or UST system in accordance with applicable requirements in 40 CFR part 280 or applicable state requirements in those states that have been delegated authority by EPA to administer the UST program pursuant to 42 U.S.C. 6991c and 40 CFR part 281.

Part 281—Approval of State Underground Storage Tank Programs

1. The authority citation for part 281 is revised to read as follows:

Authority: 42 U.S.C. 6912, 6991 (c), (d), (e), (g).

Subpart C—[Amended]

2. Section 281.39 is added to subpart C to read as follows:

§281.39 Lender liability

(a) A state program that contains a security interest exemption will be considered to be no less stringent than, and as broad in scope as, the federal program provided that the state's exemption:
   (1) Mirrors the security interest exemption provided for in 40 CFR part 280, subpart I; or
   (2) Achieves the same effect as provided by the following key criteria:
      (i) A holder, meaning a person who maintains indicia of ownership primarily to protect a security interest in a petroleum UST or UST system or facility or property on which a petroleum UST or UST system is located, who does not participate in the management of the UST or UST system as defined under §280.210 of this chapter, and who does not engage in petroleum production, refining, and marketing as defined under §280.200(b) of this chapter is not;
         (A) An "owner" of a petroleum UST or UST system or facility or property on which a petroleum UST or UST system is located for purposes of compliance with the requirements of 40 CFR part 280; or
         (B) An "operator" of a petroleum UST or UST system for purposes of compliance with the requirements of 40 CFR part 280, provided the holder is not in control of or does not have responsibility for the daily operation of the UST or UST system.
      (ii) [Reserved]
(b) [Reserved]

Appendix B

Special Contractor Prequalification Questionnaire

Date of Questionnaire _____ 19_____

_____ ( ) An Individual
Legal Name of Firm   ( ) A Partnership
                                            ( ) A Corporation
                                            ( ) A Joint Venture
                                            ( ) Other (explain)

Address   _____

                _____

City and State _____
                                        ZIP Code

Telephone   (____) _____

Description of Process

1. Contractors who wish to submit bids for either or both contracts must be successfully prequalified prior to submitting bids. To apply for prequalification, contractors must furnish the information requested in the accompanying Prequalification Questionnaire.

2. Prequalification for these contracts will be administered by a Prequalification Committee. The Committee will evaluate the completed questionnaires and comments from the references provided, and either prequalify or deny prequalification to the prospective bidder.

3. The Committee may, at its discretion, request additional information or an interview prior to making its decision to prequalify or deny prequalification.
4. The Committee will advise the prospective bidder of its decision by certified mail, stating whether or not the prospective bidder is prequalified to bid.
5. The prospective bidder may appeal a decision of the Committee. Requests for appeals must be made in writing within ten calendar days from the date of the Committee's letter. Direct the appeal to the Chief Engineer. The appeal will be heard by the Chief Engineer or his designee.
6. The decision of the Chief Engineer or his designee will be final and binding and no further appeals will be entertained.
7. Nothing in this process shall be construed as depriving the Department of the right to reject any bid prior to the award of a contract when other circumstances or developments in the opinion of the Department alter the prequalification or responsibility of the bidder.

Criteria for Prequalification

1. The Department is interested in evaluating the prospective bidder's ability to satisfactorily complete the work. The prequalification questionnaire is divided into eleven sections, A through K.
   a. Identification
   b. Licensing
   c. Experience
   d. Project Team
   e. Technical
   f. References
   g. Financial Structure
   h. Backlog
   i. Foreign Ownership
   j. Additional Information
   k. Signatures
2. Prospective bidders are invited to submit information in each of these areas to demonstrate their ability to meet the minimum prequalification requirements and their ability to perform the construction. The primary criteria for prequalification will be successful completion of past projects of a similar nature and the demonstrated capability to complete this project.
3. The project contains several key areas of work. The Department will evaluate the ability of prospective bidders to perform this work.
   For the approach spans contract, the key areas are:
   a. Major bridge construction over/in water.
   b. Precast concrete segmental bridges.
   c. Construction of major structural steel bridges.
4. The following sections will provide a general description of the various con-

struction items and the minimum criteria for prequalification. The minimum criteria listed below are necessary for successful prequalification, but are not sufficient alone for prequalification. All criteria will be considered in making the final decision to prequalify or deny prequalification.

a. Bascule Span Superstructure

   (1) General Description: Construction of a new four-lane highway movable bridge superstructure, located in the Fore River over an active navigation channel, which will remain active during the construction period. Construction will include associated structural, mechanical, electrical and control systems as required to provide a complete, operational, four-leaf, 285-feet, 6-inch trunnion-to-trunnion, bascule facility. The structure will be comprised of a steel girder and floor system of welded and bolted construction. The movable spans will be operated through mechanical gear-reduction units, powered by electric motors, and controlled by a programmable logic controller system. Completion of the upper segments of two reinforced concrete supporting piers is included in the required construction. The preliminary construction estimate for the bascule span superstructure is $35,000,000.

   (2) Minimum Prequalification Requirements: Demonstrate successful completion of one movable bridge span superstructure over an active commercial waterway within the last ten years.

b. Approach Spans: Contractors will be permitted to bid on any of the alternate approach spans contracts with the contract awards, or award, going to the lowest bid combination, or single bid, which constructs the entire superstructure and substructure.

   (1) Steel Alternative Approaches

      (a) General Description: Construction of a new, four-lane highway bridge approach superstructure, partially located over the Fore River. The main roadway structure will be approximately 4,400 feet in length, by 88 feet wide, with an additional single-lane approach ramp approximately 650 feet in length. The structure will be comprised of steel girders and bracing, of welded and bolted construction, with a cast-in-place reinforced concrete deck roadway.

      The preliminary construction estimates for the steel alternative to construct the approach spans are as follows:

- Substructure for Steel Alternate $15,000,000
- South Superstructure for Steel Alternate $25,000,000
- North Superstructure for Steel Alternate $20,000,000
- Entire Superstructure for Steel Alternate $45,000,000
- Superstructure and Substructure for Steel Alternate $60,000,000

(b) Minimum Prequalification Requirements: Demonstrate successful completion within the last five years of at least one project having a steel bridge superstructure and cast-in-place concrete deck 1,500 feet long, with multiple spans of at least 150 feet in length.

(2) Concrete Alternative Approach Superstructure

(a) General Description: Construction of a new, four-lane highway bridge approach superstructure partially located over the Fore River. The main roadway structure will be approximately 4,400 feet in length, with an additional single-lane approach ramp approximately 650 feet in length. The structure will be comprised predominantly of precast, post-tensioned, segmental concrete box girders. The majority of the superstructure will be erected using balanced cantilever methods, with some additional use of erection on falsework.

The preliminary construction estimates for the concrete alternative to construct the approach spans are as follows:

- Substructure for Concrete Alternate $15,000,000
- Superstructure for Concrete Alternate $45,000,000
- Superstructure and Substructure for
  Concrete Alternate $60,000,000

(b) Minimum Prequalification Requirements: Demonstrate successful completion within the last five years of at least one project using precast, post-tensioned, segmental, concrete box girders, 3,000 feet long, using balanced cantilever erection.

Grounds for Denial of Prequalification

Any one or more of the following may be considered as sufficient grounds for denial of prequalification of a prospective bidder if they were caused by conditions within the control of the prospective bidder.

1. More than one application submitted from an individual, firm or corporation under the same or different names, excluding joint ventures.

2. Unsatisfactory performance on past work for the Department based upon information available to the Prequalification Committee.

3. Incomplete work in progress for the Department which might be further delayed, or other incomplete work which might hinder or prevent the prompt completion of this project.

4. Deceptive or fraudulent statements either on the Prequalification Questionnaire or during any interviews before the Prequalification Committee.

5. Unreasonable refusal to pay all bills for labor and materials on any former or existing contracts.

6. Failure to satisfy one or more judgments relating to any previous contracts.
7. Insufficient experience, resources or financial capability to successfully execute and complete the contract.

The above is intended to be only a partial list of possible grounds for denial of prequalification. The Prequalification Committee will consider all information presented in making its decision to grant or deny prequalification.

Instructions for Completing Prequalification Questionnaire

The following are instructions for completing the Special Prequalification Process Questionnaire on pages Q-1 through Q-8:

1. Answer all questions and provide all information in this Questionnaire. Partial applications will not be accepted and will be returned to the prospective bidder with no action taken.
2. In the case of a joint venture, provide for each member of the joint venture the information required in all questionnaire items. If a joint venture is successfully prequalified, bids can only be accepted for the joint venture. Individual members of a prequalified joint venture may not submit bids on their own unless they individually apply for and receive prequalified status.
3. Submit six copies of the reply to this questionnaire.

Special Prequalification Process Questionnaire

A. Identification
   1. What business entity has the prospective bidder adopted (i.e., corporation, joint venture, partnership, or sole proprietorship)?
   2. How many years has this entity been in business as a contractor?
   3. If a corporation, provide the following:
      a. Name under which the entity would bid.
      b. State of incorporation and date.
      c. Home office address, telephone and facsimile number.
      d. Names of all officers and directors.
      e. Name of principal who will represent the entity in any dealings with regard to this prequalification.
      f. Complete address, if any, of prospective bidder's local office in the State of Maine.
   4. If a joint venture or partnership, provide the following:
      a. Name under which the entity would bid.
      b. Date established.
      c. Home office address, telephone and facsimile number of the entity. In the event the prospective bidder is a joint venture, provide address and telephone numbers for each member of the joint venture.
      d. What type of partnership the entity is, general or limited.

e. Name of the principal who will represent the prospective bidder in any dealings with regard to this prequalification.
   f. Identify all the principals of the prospective bidder's organization.
B. Licensing
   1. List jurisdictions and trade categories in which your organization is legally qualified to do business, and indicate registration or license numbers, if applicable.
   2. List jurisdictions in which your organization's partnership or trade name is filed.
   3. List jurisdiction(s) and explain if any licenses have been rescinded or revoked.
C. Experience
   1. Background Experience: List bridge or similar construction projects on which the entity has worked during the past ten years.
      a. Owner, project name, location.
      b. Owner's contract number.
      c. Name, address, and phone number of owner's project manager or other reference familiar with the project and your role in it.
      d. Original contract value and final contract value.
      e. Contract start date, initial contract completion date, and actual contract completion date.
      f. A technical description in narrative form of the project and the portion of the project the entity began and completed.
   2. State the average annual dollar volume of construction work performed during the past five years. Indicate the name, location, and dollar value of each project completed.
   3. Detailed Experience: Provide a detailed factual description of three recently completed bridge projects which you feel are similar in type and size to the Portland Bridge. Emphasis should be placed on the technical complexity, approach, and your major responsibilities in each project.
   4. Termination:
      a. Indicate if the prospective bidder or elements of the prospective bidder have ever had a contract terminated for any reason.
      b. If yes, identify the project, the owner, the owner's contract number, and explain the circumstances concerning the termination.
   5. Claims and Suits: For each of the projects listed in (1) and (3) above, indicate the following:
      a. Which projects, if any, the prospective bidder failed to complete and the reason the project was not completed.
      b. Explain the nature and circumstances of any judgments, claims, arbitration proceedings or suits pending or outstanding against your organization or its officers.
      c. Any suits filed or arbitration requested by or against the prospective bidder with regard to a construction contract within the last five years.
D. Project Team
   1. Organization:

a. Provide an organization chart showing how you will organize your work force for this project. Indicate the titles of the key project personnel. Include, if applicable, the project executive, project manager, construction manager, project superintendent, project engineer, project scheduler, cost engineer, safety engineer, procurement manager. Identify by area of responsibility, each superintendent and field engineer. Include both on-site and home office personnel.
b. Describe how the project organization would relate to your home office organization(s).
2. Personnel:
a. Identify which positions you plan to staff from your experienced employees and which from new employees to be hired for this project.
b. Provide resumes of the top key personnel identified in the organization chart. Specifically identify:
(1) Position and name
(2) Education
(3) Years of construction experience
(4) Years with the firm, or firm making up part of a joint venture
(5) Registrations, awards, publications
(6) Experience—highlight experience similar to that anticipated on this project
3. List and explain any present commitments of the individuals listed in (2) above.
4. Construction Engineer: If construction engineering services are planned, identify the individual or firm who will be performing these construction engineering services. If a separate engineering firm is being used, provide the same detailed information requested in paragraph 2 above for the key individuals who will be assigned to this project.
E. Technical Section
1. Construction Schedule:
a. Provide a brief narrative of how the prospective bidder would schedule the project. Include a discussion of the process to be used over the duration of the project.
b. Provide a brief history of the prospective bidder's experience and capabilities in scheduling.
c. Identify the scheduling software and computer hardware you use.
d. Describe any cost loaded schedule payment procedure you have used. Identify the projects.
e. Enclose an example of a typical schedule with update and analysis report, used by the prospective bidder on a recent project. If possible, the project should be similar in size and complexity to the Portland–South Portland Bridge.
2. Quality Control: Provide a brief narrative describing how the prospective bidder will ensure the highest quality of the project. The narrative should include:
a. A brief statement of company policy as to quality control.

b. A brief description of the company's quality control organization on previous projects, to include the functional relationships, identity and responsibilities of quality control personnel;
3. Safety:
   a. Provide a brief narrative describing how the entity would ensure worker safety on the project. Include a description of anticipated safety staffing, and a skeletal outline of a proposed safety program.
   b. Use your last year's OSHA No. 200 Log to indicate (if a joint venture, provide the information for each member):
      (1) Number of lost workday cases
      (2) Number of restricted work cases
      (3) Number of cases with medical attention only
      (4) Number of fatalities
   c. Employee hours worked last year. Include all field construction employees on your payroll.
   d. List the prospective bidder's Insurance Experience Modification Rate for the three most recent years. If a joint venture, list for each member of the venture.
4. Partnering: The successful bidder will be asked to participate in a voluntary partnering program. Provide a brief history of the prospective bidder's experience with partnering.

F. References
   1. Trade References: List at least six trade references with name of contact person and complete address and telephone number. List three suppliers, two subcontractors, and one trade organization such as the AGC.
   2. Bank References: Include name of contact person and complete address and telephone number of the bank(s) which hold your business accounts.

G. Financial Structure
   1. Support Documents: Submit the following documents in support of the prospective bidder's financial structure.
      a. Audited financial statements for the last three years with accompanying notes and auditor's reports by an independent CPA. Identify separately financial statements for construction activities.
      b. Attach your organization's latest balance sheet and income statement showing the following items:
         (1) Current Assets (e.g., cash, joint venture accounts, accounts receivable, notes receivable, accrued income, deposits, materials inventory and prepaid expenses).
         (2) Fixed Assets (gross) and related accumulated depreciation.
         (3) Other Assets.
         (4) Current Liabilities (e.g., accounts payable, notes payable, accrued expenses, provision for income taxes, accrued salaries and accrued payroll taxes).
         (5) Other Liabilities and Equity (e.g., long-term debt, capital stock, additional paid in capital, authorized and outstanding shares par values, earned surplus and retained earnings).

(6) Name, address, and contact person of accounting firm preparing attached financial statements, and date thereof.
(7) Is the attached financial statement of the identical organization named on page one?
(8) If the answer to (7) above is no, explain the relationship and financial responsibility of the organization whose financial statement is provided (e.g., parent–subsidiary).
(9) Will the organization whose financial statement is attached act as a guarantor of the contract for construction?
(10) Is the prospective bidder's organization, its parent company(s), subsidiary(s) or partner(s) currently involved in any litigation not noted in the accompanying footnotes to the audited financial statements? If yes, explain.
c. Name of prospective bidder's bonding company.
d. Provide the name, address, and telephone number of the bonding company's agent.
2. If the entity is a joint venture, provide the following information:
a. Copy of Joint Venture Agreement or Letter of Commitment.
b. The responsibility of each member of the joint venture.
c. Structure of the joint venture.

H. Backlog
1. Provide a list of current contracts now underway and a list of projects for which the entity is now competing or negotiating. Include the following information:
a. Name of projects, owners, and locations.
b. Value of contracts.
c. Incomplete backlog.
d. Brief description of type of work.
e. Duration of work (start date and estimated completion date).

I. Foreign Ownership
1. Is the prospective bidder, or if a joint venture or partnership, any member or partner, wholly or partially owned by a foreign entity? Foreign is intended to mean relating to, belonging to, or owing allegiance to a country other than the United States. If yes, what foreign country or nationality?
2. If the answer to (1) was yes, answer the following questions.
a. What is the name of the foreign entity?
b. What percentage of ownership is controlled by the foreign entity?
c. What percentage of management, engineering or administrative personnel within your company headquarters or divisional headquarters are U.S. citizens?

J. Additional Information
1. Provide here any additional information which has not been included in this questionnaire and which the entity feels would contribute to the prequalification review.

K. Signatures
1. The signatory of this questionnaire warrants that all statements made herein are accurate and complete to the best of his (her) knowledge, information and belief.
2. The undersigned authorize(s) and request(s) any public official, engineer, architect, surety company, bank, depository, material or equipment manufacturer or distributor or any person, firm, or corporation to furnish to the Maine Department of Transportation, any information the Department deems necessary to verify the statements made herein to evaluate this prequalification questionnaire. None of the furnished information will be distributed outside the Department unless required by law.

Dated at _____ this _____ day of_____, 19____.

_____ By _____
Name of Organization

_____
Title of Person Signing
(If Corporate, Give Seal)

STATE OF _____

COUNTY OF_____

Before me personally appeared _____ known to me to be the person described in and who executed the foregoing instrument.

WITNESS my hand and official seal, this _____ day of _____ A.D. 19____.
My commission expires:

_____
Notary Public

State of _____

_____

Appendix C

OSHA Regulations: Hazard Communications Standard

Part Number: __1910__

Standard Number: __1910.1200__

Title: __Hazard Communications Standard__

(a) "Purpose"

(1) The purpose of this section is to ensure that the hazards of all chemicals produced or imported are evaluated, and that information concerning their hazards is transmitted to employers and employees. This transmittal of information is to be accomplished by means of comprehensive hazard communications programs, which are to include container labeling and other forms of warning, material safety data sheets and employee training.

(2) This occupational safety and health standard is intended to address comprehensively the issue of evaluating the potential hazards of chemicals, and communicating information concerning hazards and appropriate protective measures to employees, and to preempt any legal requirements of State, or political subdivisions of a State, pertaining to this subject. Evaluating the potential hazards of chemicals, and communicating information concerning hazards and appropriate protective measures to employees, may include, for example, but is not limited to, provisions for: developing and maintaining a written hazard communication program for the workplace, including lists of hazardous chemicals present; labeling of containers of chemicals in the workplace, as well as of containers of chemicals being shipped to other employees and downstream employers; and

development and implementation of employee training programs regarding hazards of chemicals and protective measures. Under Section 18 of the Act, no State or political subdivision of a State may adopt or enforce, through any court or agency, any requirement relating to the issue addressed by this Federal standard, except pursuant to a Federally approved State plan.

(b) "Scope and Applications"

(1) This section requires chemical manufacturers or importers to assess the hazards of chemicals which they produce or import, and all employers to provide information to their employees about the hazardous chemicals to which they are exposed, by means of a hazard communication program, labels and other forms of warning, material safety data sheets, and information and training. In addition, this section requires distributors to transmit the required information to employers. (Employers who do not produce or import chemicals need only focus on those parts of this rule that deal with establishing a workplace program and communicating information to their workers. Appendix E of this section is a general guide for such employers to help them determine their compliance obligations under the rule.)

(2) This section applies to any chemical which is known to be present in the workplace in such a manner that employees may be exposed under normal conditions of use or in a foreseeable emergency.

(3) This section applies to laboratories only as follows:
  (i) Employers shall ensure that labels on incoming containers of hazardous chemicals are not removed or defaced;
  (ii) Employers shall maintain any material safety data sheets that are received with incoming shipments of hazardous chemicals, and ensure that they are readily accessible during each work shift to laboratory employees when they are in their work areas;
  (iii) Employers shall ensure that laboratory employees are provided information and training in accordance with paragraph (h) of this section, except for the location and availability of the written hazard communication program under paragraph (h) (2) (iii) of this section; and,
  (iv) Laboratory employers that ship hazardous chemicals are considered to be either a chemical manufacturer or a distributor under this rule, and thus, must ensure that any containers of hazardous chemicals leaving the laboratory are labeled in accordance with paragraph (f)(1) of this section, and that a material safety data sheet is provided to distributors and other employers in accordance with paragraphs (g) (6) and (g) (7) of this section.

(4) In work operations where employees only handle chemicals in sealed containers which are not opened under normal conditions of use (such as are found in marine cargo handling, warehousing, or retail sales), this section applies to these operations only as follows:

(i) Employers shall ensure that labels on incoming containers of hazardous chemicals are not removed or defaced;

(ii) Employers shall maintain copies of any material safety data sheets that are received with incoming shipments of the sealed containers of hazardous chemicals, shall obtain a material safety data sheet as soon as possible for sealed containers of hazardous chemicals received without a material safety data sheet if an employee requests the material safety data sheet, and shall ensure that the material safety data sheets are readily accessible during each work shift to employees when they are in their work area(s); and,

(iii) Employers shall ensure that employees are provided with information and training in accordance with paragraph (h) of this section (except for the location and availability of the written hazard communication program under paragraph (h) (2) (iii) of this section), to the extent necessary to protect them in the event of a spill or leak of a hazardous chemical from a sealed container.

(5) This section does not require labeling of the following chemicals:

(i) Any pesticide as such term is defined in the Federal Insecticide, Fungicide, and Rodenticide Act (7 U.S.C. 136 et seq.) when subject to the labeling requirements of that Act and labeling regulations issued under that Act by the Environmental Protection Agency;

(ii) Any chemical substance or mixture as such terms are defined in the Toxic Substances Control Act (15 U.S.C. 2601 et seq.), when subject to the labeling requirements of that Act and labeling regulations issued under that Act by the Environmental Protection Agency;

(iii) Any food, food additive, color additive, drug, cosmetic, or medical or veterinary device or product, including materials intended for use as ingredients in such products (e.g., flavors and fragrances) as such terms are defined in the Federal Food, Drug, and Cosmetic Act (21 U.S.C. 301 et seq.) or the Virus-Serum-Toxin Act of 1913 (21 U.S.C. 151 et seq.), and regulations issued under those Acts, when they are subject to the labeling requirements under those Acts by either the Food and Drug Administration or the Department of Agriculture;

(iv) Any distilled spirits (beverage alcohols), wine, or malt beverage intended for nonindustrial use, as such terms are defined in the Federal Alcohol Administration Act (27 U.S.C. et seq.) and regulations issued under that Act, when subject to the labeling requirements of that Act and labeling regulations issued under that Act by the Bureau of Alcohol, Tobacco, and Firearms;

(v) Any consumer product or hazardous substance as those terms are defined in the Consumer Product Safety Act (15 U.S.C. 2051 et seq.) and Federal Hazardous Substances Act (15 U.S.C. 1261 et seq.), respectively, when subject to a consumer product safety standard or labeling requirements of those Acts, or regulations issued under those Acts by the Consumer Product Safety Commission; and,

(vi) Agricultural or vegetable seed treated with pesticides and labeled in

accordance with the Federal Seed Act (7 U.S.C. 1551 et seq.) and the labeling regulations issued under that Act by the Department of Agriculture.
(6) This section does not apply to:
  (i) Any hazardous waste as such term is defined by the Solid Waste Disposal Act, as amended by the Resource Conservation and Recovery Act of 1976, as amended (42 U.S.C. 6901 et seq.), when subject to regulations issued under that Act by the Environmental Protection Agency;
  (ii) Any hazardous substance as such term is defined by the Comprehensive Environmental Response, Compensation and Liability Act (CERCLA) (42 U.S.C. 9601 et seq.) when the hazardous substance is the focus of remedial or removal action being conducted under CERCLA in accordance with the Environmental Protection Agency regulations;
  (iii) Tobacco or tobacco products;
  (iv) Wood or wood products, including lumber which will not be processed, where the chemical manufacturer or importer can establish the only hazard they pose to employees is the potential for flammability or combustibility (wood or wood products which have been treated with a hazardous chemical covered by this standard, and wood which may be subsequently sawed or cut, generating dust, are not exempted);
  (v) Articles (as that term is defined in paragraph (c) of this section);
  (vi) Food or alcohol beverages which are sold, used, or prepared in a retail establishment (such as a grocery store, restaurant, or drinking place), and foods intended for personal consumption by employees while in the workplace;
  (vii) Any drug, as that term is defined in the Federal Food, Drug, and Cosmetic Act (21 U.S.C. 301 et seq.), when it is in solid, final form for direct administration to the patient (e.g., tablets or pills); drugs which are packaged by the chemical manufacturer for sale to consumers in a retail establishment (e.g., over-the-counter drugs); and drugs intended for personal consumption by employees while in the workplace (e.g., first aid supplies);
  (viii) Cosmetics which are packaged for sale to consumers in a retail establishment, and cosmetics intended for personal consumption by employees while in the workplace;
  (ix) Any consumer product or hazardous substance, as those terms are defined in the Consumer Product Safety Act (15 U.S.C. 2051 et seq.) and Federal Hazardous Substances Act (15 U.S.C. 1261 et seq.), respectively, where the employer can show that it is used in the workplace for the purpose intended by the chemical manufacturer or importer of the product, and the use results in a duration and frequency of exposure which is not greater than the range of exposures that could reasonably be experienced by consumers when used for the purpose intended;

(x) Nuisance particulars where the chemical manufacturer or importer can establish that they do not pose any physical or health hazard covered under this section;
(xi) Ionizing and nonionizing radiation; and
(xii) Biological hazards.

(c) "Definitions"

"Article" means a manufactured item other than a fluid or particle:

(i) which is formed to a specific shape or design during manufacture;
(ii) which has end use function(s) dependent in whole or in part upon its shape or design during end use; and
(iii) which, under normal conditions of use, does not pose a physical hazard or health risk to employees.

"Assistant Secretary" means the Assistant Secretary of Labor for Occupational Safety and Health, U.S. Department of Labor, or designee.

"Chemical" means any element, chemical compound, or mixture of elements and/or compounds.

"Chemical manufacturer" means an employer with a workplace where chemical(s) are produced for use or distribution.

"Chemical name" means the scientific designation of a chemical in accordance with the nomenclature system developed by the International Union of Pure and Applied Chemistry (IUPAC) or the Chemical Abstracts Service (CAS) rules of nomenclature, or a name which will clearly identify the chemical for the purpose of conducting a hazard evaluation.

"Combustible liquid" means any liquid having a flashpoint at or above 100 deg. F (37.8 deg. C), but below 200 deg. F (93.3 deg. C), except any mixture having components with flashpoints of 200 deg. F (93.3 deg. C), or higher, the total volume of which make up 99 percent or more of the total volume of the mixture.

"Commercial account" means an arrangement whereby a retail distributor seeks hazardous chemicals. To an employer, generally in large quantities over time and/or at costs that are below the regular retail price.

"Common name" means any designation or identification such as code name, code number, trade name, brand name, or general name used to identify a chemical other than by its chemical name.

"Compressed gas" means:

(i) A gas mixture of gases having, in a container, an absolute pressure exceeding 40 psi at 70 deg. F (21.1 deg. C);
(ii) A gas or mixture of gases having, in a container, an absolute pressure

exceeding 104 psi at 130 deg. F (54.4 deg. C) regardless of the pressure at 70 deg. F (21.1 deg. C); or

(iii) A liquid having a vapor pressure exceeding 40 psi at 100 deg. F (37.8 deg. C) as determined by ASTM D-323-72.

"Container" means any bag, barrel, bottle, box, can, cylinder, drum, reaction vessel, storage tank, or the like that contains a hazardous chemical. For purposes of this section, pipes or piping systems, and engines, fuel tanks, or other operating systems in a vehicle, are not considered to be containers.

"Designated representative" means any individual or organization to whom an employee gives written authorization to exercise such employee's rights under this section. A recognized or certified collective bargaining agent shall be treated automatically as a designated representative without regard to written employee authorization.

"Director" means the Director, National Institute for Occupational Safety and Health, U.S. Department of Health and Human Services, or designee.

"Distributor" means a business, other than a chemical manufacturer or importer, which supplies hazardous chemicals to other distributors or to employers.

"Employee" means a worker who may be exposed to hazardous chemicals under normal operating conditions or in foreseeable emergencies. Workers such as office workers or bank tellers who encounter hazardous chemicals only in nonroutine, isolated instances are not covered.

"Employer" means a person engaged in a business where chemicals are either used, distributed, or are produced for use or distribution, including a contractor or subcontractor.

"Explosive" means a chemical that causes a sudden, almost instantaneous release of pressure, gas, and heat when subjected to sudden shock, pressure, or high temperature.

"Exposure or exposed" means that an employee is subjected in the course of employment to a chemical that is a physical or health hazard, and includes potential (e.g., accidental or possible) exposure. "*Subjected*" in terms of health hazards includes any route of entry (e.g., inhalation, ingestion, skin contact or absorption).

"Flammable" means a chemical that falls into one of the following categories:

(i) "Aerosol, flammable" means an aerosol that, when tested by the method described in 16 CFR 1500.45, yields a flame projection exceeding 18 inches at full valve opening, or a flashback (a flame extending back to the valve) at any degree of valve opening;

(ii) "Gas, flammable" means

(a) A gas that, at ambient temperature and pressure, forms a flammable mixture with air at a concentration of thirteen (13) percent by volume or less, or;

(b) A gas that, at ambient temperature and pressure, forms a range of flammable mixtures with air wider than twelve (12) percent of volume, regardless of the lower limit;

(iii) "Liquid, flammable" means any liquid having a flashpoint below 100 deg. F (37.8 deg. C), except any mixture having components with flashpoints of 100 deg. F (37.8 deg. C) or higher, the total of which make up 99 percent or more of the total volume of the mixture.

(iv) "Solid, flammable" means a solid, other than a blasting agent or explosive as defined in 1910.109 (a), that is liable to cause fire through friction, absorption of moisture, spontaneous chemical change, or retained heat from manufacturing or processing, or which can be ignited readily and when ignited burns so vigorously and persistently as to create a serious hazard. A chemical shall be considered to be a flammable solid if, when tested by the method described in 16 CFR 1500.44, it ignites and burns with a self-sustained flame at a rate greater than one-tenth of an inch per second along its major axis.

"Flashpoint" means a temperature at which a liquid gives off a vapor in sufficient concentration to ignite when tested as follows:

(i) Tagliabue Closed Tester (see American National Standard Method of Test for Flash Point by Tag Closed Tester, Z11.24-1979 (ASTM D 56-79)) for liquids with a viscosity of less than 45 Saybolt Universal Seconds (SUS) at 100 deg. F (37.8 deg. C), that do not contain suspended solids and do not have tendency to form a surface film under test; or

(ii) Pensky-Martens Closed Tester (see American National Standard Method of Test for Flash Point by Pensky-Martens Closed Tester, Z11.7-1979 (ASTM D 93-79)) for liquids with a viscosity equal to or greater than 45 SUS at 100 deg. F (37.8 deg. C), that contain suspended solids, or that have a tendency to form a surface film under test; or

(iii) Setaflash Closed Tester (see American National Standard Method of Test for Flash Point by Setaflash Closed Tester (ASTM D 3278-78)).

Organic peroxides, which undergo auto-accelerating thermal decomposition, are excluded from any of the flashpoint determination methods specified above.

"Foreseeable emergency" means any potential occurrence such as, but not limited to, equipment failure, rupture of containers, or failure of control equipment which could result in an uncontrolled release of a hazardous chemical into the workplace.

"Hazardous chemical" means any words, pictures, symbols, or combination thereof appearing on a label or other appropriate form of warning which convey the specific physical and health hazard(s), including target organ effects, of the chemical(s) in the container(s). (See the definitions for "physical hazard" to determine the hazards which must be covered.)

"Health hazard" means a chemical for which there is statistically significant evidence based on at least one study conducted in accordance with established scientific principles that acute or chronic health effects may occur in exposed employees. The term "health hazard" includes chemicals which are carcinogens, toxic or highly toxic agents, reproductive toxins, irritants, corrosive, sensitizers, hepatoxins, nephotoxins, neurotoxins, agents which act on the hematopoietic system, and agents which damage the lungs, skin, eyes, or mucous membranes. Appendix A provides further definitions and explanations of the scope of health hazards covered by this section, and Appendix B describes the criteria to be used to determine whether or not a chemical is to be considered hazardous for purposes of this standard.

"Identity" means any chemical or common name which is indicated on the material safety data sheet (MSDS) for the chemical. The identity used shall permit cross-references to be made among the required list of hazardous chemicals, the label and the MSDS.

"Immediate use" means that the hazardous chemical will be under the control of and used only by the person who transfers it from a labeled container and only within the work shift in which it is transferred.

"Importer" means the first business with employees within the Customs Territory of the United States which receives hazardous chemicals produced in other countries for the purpose of supplying them to distributors or employers within the United States.

"Label" means any written, printed, or graphic material displayed on or affixed to containers of hazardous chemicals.

"Material safety data sheet" (MSDS) means written or printed material concerning a hazardous chemical which is prepared in accordance with paragraph (g) of this section.

"Mixture" means any combination of two or more chemicals if the combination is not, in whole or in part, the result of a chemical reaction.

"Organic peroxide" means an organic compound that contains the bivalent -O-O-structure and which may be considered to be a structural deviative of hydrogen peroxide where one or both of the hydrogen atoms has been replaced by an organic radical.

"Oxidizer" means a chemical other than a blasting agent or explosive as defined in 1910.109(a), that initiates or promotes combustion in other materials, thereby causing fire either of itself or through the release of oxygen or other gases.

"Physical hazard" means a chemical for which there is scientifically valid evidence that it is a combustible liquid, a compressed gas, flammable, an organic peroxide, an oxidizer, pyrophoric, unstable (reactive) or water-reactive.

"Produce" means to manufacture, process, formulate, blend, extract, generate, emit, or repackage.

"Pyrophocric" means a chemical that will ignite spontaneously in air at a temperature of 130 deg. F (54.4 deg. C) or below.

"Responsible party" means someone who can provide additional information on the hazardous chemical and appropriate emergency procedures, if necessary.

"Specific chemical identity" means the chemical name, Chemical Abstracts Service (CAS) Registry Number, or any other information that reveals the precise chemical designation of the substance.

"Trade secret" means any confidential formula, pattern, process, device, information or compilation of information that is used in an employer's business, and that gives the employer an opportunity to obtain an advantage over competitors who do not know or use it. Appendix D sets out the criteria to be used in evaluating trade secrets.

"Unstable (reactive)" means a chemical which in the pure state, or as produced or transported, will vigorously polymerize, decompose, condense, or will become self-reactive under conditions of shocks, pressure or temperature.

"Use" means to package, handle, react, emit, extract, generate as a by-product, or transfer.

"Water-reactive" means a chemical that reacts with water to release a gas that is either flammable or presents a health hazard.

"Work area" means a room or defined space in a workplace where hazardous chemicals are produced or used, and where employees are present.

"Workplace" means an establishment, job site, or project, at one geographical location containing one or more work areas.

(d) "Hazard Determination"

(1) Chemical manufacturers and importers shall evaluate chemicals produced in their workplaces or imported by them to determine if they are hazardous. Employers are not required to evaluate chemicals unless they choose not to rely on the evaluation performed by the chemical manufacturer or importer for the chemical to satisfy this requirement.

(2) Chemical manufacturers, importers or employers evaluating chemicals shall identify and consider the available scientific evidence concerning such hazards. For health hazards, evidence which is statistically significant and which is based on at least one positive study conducted in accordance with established scientific principles is considered to be sufficient to establish a hazardous effect if the results of the study meet the definitions of health hazards in this section. Appendix A shall be consulted for the scope of health hazards covered, and Appendix B shall be consulted for the criteria to be followed with respect to the completeness of the evaluation, and the data to be reported.

(3) The chemical manufacturer, importer or employer evaluating chemicals shall treat the following sources as establishing that the chemicals listed in them are hazardous:

(i) 29 CFR part 19110, subpart Z, Toxic and Hazardous Substances, Occupational Safety and Health Administration (OSHA); or,

(ii) Threshold Limit Values for Chemical Substances and Physical Agents in the Work Environment, American Conference of Governmental Industrial Hygienists (ACGIH) (latest edition). The chemical manufacturer, importer or employer is still responsible for evaluating the hazards associated with the chemicals in these source lists in accordance with the requirements of this standard.

(4) Chemical manufacturers, importers and employers evaluating chemicals shall treat the following sources as establishing that a chemical is a carcinogen or potential carcinogen for hazard communication purposes:

(i) National Toxicology Program (NTP), "Annual Report on Carcinogens" (latest editions); or,

(ii) International Agency for Research on Cancer (IARC) "Monographs" (latest editions); or

(iii) 29 CFR part 1910, subpart Z, Toxic and Hazardous Substances, Occupational Safety and Health Administration.

Note: The "Registry of Toxic Effects of Chemical Substances" published by the National Institute for Occupational Safety and Health indicates whether a chemical has been found by NTP or IARC to be a potential carcinogen.

(5) The chemical manufacturer, importer, or employer shall determine the hazards of mixtures of chemicals as follows:

(i) If a mixture has been tested as a whole to determine its hazards, the results of such testing shall be used to determine whether the mixture is hazardous;

(ii) If a mixture has not been tested as a whole to determine whether the mixture is a health hazard, the mixture shall be assumed to present the same health hazards as do the components which comprise one percent (by weight or volume) or greater of the mixture, except that the mixture shall be assumed to present a carcinogenic hazard if it contains a component in concentrations of 0.1 percent or greater which is considered to be a carcinogen under paragraph (d)(4) of this section;

(iii) If a mixture has not been tested as a whole to determine whether the mixture is a physical hazard, the chemical manufacturer, importer, or employer may use whatever scientifically valid data is available to evaluate the physical hazard potential of the mixture; and

(iv) If the chemical manufacturer, importer, or employer has evidence to indicate that a component present in the mixture in concentrations of less than one percent (or in the case of carcinogens, less than 0.1 percent) could be released in concentrations which would exceed an established OSHA permissible exposure limit or ACGIH Threshold Limit Value, or could present a health risk to employees in those concentrations, the mixture shall be assumed to present the same hazard.

(6) Chemical manufacturers, importers, or employers evaluating chemicals shall describe in writing the procedures they use to determine the hazards of the chemical they evaluate. The written procedures are to be made available, upon request, to employees, their designated representatives, the Assistant Secretary and the Director. The written description may be incorporated into the written hazard communication program required under paragraph (e) of this section.

(e) "Written Hazard Communication Programs"

(1) Employers shall develop, implement, and maintain at each workplace, a written hazard communication program which at least describes how the criteria specified in paragraphs (f), (g) and (h) of this section for labels and other forms of warning, material safety data sheets, and employee information and training will be met, and which also includes the following:
    (i) A list of the hazardous chemicals known to be present using an identity that is referenced on the appropriate material safety data sheet (the list may be compiled for the workplace as a whole or for individual work areas); and,
    (ii) The methods the employer will use to inform employees of the hazards of nonroutine tasks (for example, the cleaning of reactor vessels), and the hazards associated with chemicals contained in unlabeled pipes in their work areas.

(2) Multiemployer workplaces: Employers who produce, use, or store hazardous chemicals at a workplace in such a way that the employees of other employer(s) may be exposed (for example, employees of a construction contractor on-site) shall additionally ensure that the hazard communication programs developed and implemented under this paragraph (e) include the following:
    (i) The methods the employer will use to provide the other employer(s) on-site access to material safety data sheets for each hazardous chemical the other employer's employees may be exposed to while working;
    (ii) The methods the employer will use to inform the other employer(s) of any precautionary measures that need to be taken to protect employees during the workplace's normal operating conditions and in foreseeable emergencies; and,
    (iii) The methods the employer will use to inform the other employer(s) of the labeling system used in the workplace.

(3) The employer may rely on an existing hazard communication program to comply with these requirements, provided that it meets the criteria established in this paragraph (e).

(4) The employer shall make the written hazard communication program available, upon request, to employees, their designated representatives, the Assistant Secretary and the Director, in accordance with the requirements of 29 CFR 1910.20 (e).

(5) Where employees must travel between workplaces during a workshift, i.e., their work is carried out at more than one geographical location, the written hazard communication program may be kept at the primary workplace facility.

(f) "Labels and Other Forms of Warning"

(1) The chemical manufacturer, importer, or distributor shall ensure that each container of hazardous chemical leaving the workplace is labeled, tagged or marked with the following information:
   (i) Identity of the hazardous chemical(s);
   (ii) Appropriate hazard warnings; and
   (iii) Name and address of the chemical manufacturer, importer, or other responsible party.

(2) (i) For solid metal (such as a steel beam or a metal casting), solid wood, or plastic items that are not exempted as articles due to their downstream use, or shipments of whole grain, the required label may be transmitted to the customer at the time of the initial shipment, and need not be included with subsequent shipments to the same employer unless the information on the label changes; and
   (ii) The label may be transmitted with the initial shipment itself, or with the material safety data sheet that is to be provided prior to or at the time of the first shipment; and
   (iii) This exception to requiring labels on every container of hazardous chemicals is only for the solid material itself, and does not apply to hazardous chemicals used in conjunction with, or known to be present with, the material and to which employees handling the items in transit may be exposed (for example, cutting fluids or pesticides in grains).

(3) Chemical manufacturers, importers, or distributors shall ensure that each container of hazardous chemicals leaving the workplace is labeled, tagged, or marked in accordance with this section in a manner which does not conflict with the requirements of the Hazardous Materials Transportation Act (46 U.S.C. 1810 et seq.) and regulations issued under that Act by the Department of Transportation.

(4) If the hazardous chemical is regulated by OSHA in a substance-specific health standard, the chemical manufacturer, or importer, distributor or employer, shall ensure that the labels or other forms of warning used are in accordance with the requirements of that standard.

(5) Except as provided in paragraphs (f) (6) and (f) (7) of this section, the employer shall ensure that each container of hazardous chemicals in the workplace is labeled, tagged, or marked with the following information:
   (i) Identity of the hazardous chemical(s) contained therein; and,
   (ii) Appropriate hazard warnings, or alternatively, words, pictures, symbols, or combination thereof, which provide at least general information regarding the hazards of the chemicals, and which, in

conjunction with the other information immediately available to employees, provide specific information regarding the physical and health hazards of the hazardous chemical.

(6) The employer may use signs, placards, process sheets, batch tickets, operating procedures, or other such written materials in lieu of affixing labels to individual stationary process containers, as long as the alternative method identifies the containers to which it is applicable and conveys the information required by paragraph (f) (5) of this section to be on a label. The written materials shall be readily accessible to the employees in their work area throughout each work shift.

(7) The employer is not required to label portable containers into which hazardous chemicals are transferred from labeled containers, and which are intended only for the immediate use of the employee who performs the transfer. For purposes of this section, drugs which are dispensed by a pharmacy to a health care provider for direct administration to a patient are exempted from labeling.

(8) The employer shall not remove or deface existing labels on incoming containers of hazardous chemicals, unless the container is immediately marked with the required information.

(9) The employer shall ensure that labels or other forms of warning are legible, in English, and prominently displayed on the container, or readily available in the work area throughout each work shift. Employers having employees who speak other languages may add the information in their language to the material presented, as long as the information is presented in English as well.

(10) The chemical manufacturer, importer, distributor or employer need not affix new labels to comply with this section if existing labels already convey the required information.

(11) Chemical manufacturers, importers, distributors, or employers who become newly aware of any significant information regarding the hazards of a chemical shall revise the labels for the chemical within three months of becoming aware of the new information. Labels on containers of hazardous chemicals shipped after that time shall contain the new information. If the chemical is not currently produced or imported, the chemical manufacturer, importer, distributor, or employer shall add the information to the label before the chemical is shipped or introduced into the workplace again.

(g) "Material Safety Data Sheets"

(1) Chemical manufacturers and importers shall obtain or develop a material safety data sheet for each hazardous chemical they produce or import. Employers shall have a material safety data sheet in the workplace for each hazardous chemical which they use.

(2) Each material safety data sheet shall be in English (although the employer may maintain copies in other languages as well), and shall contain at least the following information:

(i) The use on the label, and except as provided for in paragraph (i) of this section on trade secrets:
  (A) If the hazardous chemical is a single substance, its chemical and common name (s);
  (B) If the hazardous chemical is a mixture which has been tested as a whole to determine its hazards, the chemical and common name(s) of the ingredients which contribute to these known hazards, and the common name(s) of the mixture itself; or
  (C) If the hazardous chemical is a mixture which has not been tested as a whole:
    {1} If chemical and common name(s) of all ingredients which have been determined to be health hazards, and which comprise 1% or greater of the composition, except that chemicals identified as carcinogens under paragraph (d) of this section shall be listed if the concentrations are 0.1% or greater; and
    {2} The chemical and common name(s) of all ingredients which have been determined to be health hazards, and which comprise less than 1% (0.1% for carcinogens) of the mixture, if there is evidence that the ingredient(s) could be released from the mixture in concentrations which would exceed an established OSHA permissible exposure limit or ACGIH Threshold Limit Value, or could present a health risk to employee; and,
    {3} The chemical and common name(s) of all ingredients which have been determined to present a physical hazard when present in the mixture;
(ii) Physical and chemical characteristics of the hazardous chemical (such as vapor pressure, flashpoint);
(iii) The physical hazards of the hazardous chemical, including the potential for fire, explosion, and reactivity;
(iv) The health hazards of the hazardous chemical, including signs and symptoms of exposure, and any medical conditions which are generally recognized as being aggravated by exposure to the chemical;
(v) The primary route(s) of entry;
(vi) The OSHA permissible exposure limit, ACGIH Threshold Limit Value, and any other exposure limit used or recommended by the chemical manufacturer, importer, or employer preparing the material safety data sheet, where available;
(vii) Whether the hazardous chemical is listed in the National Toxicology Program (NTP) Annual Report on Carcinogens;
(viii) Any generally applicable precautions for safe handling and use which are known to the chemical manufacturer, importer, or employer preparing the material safety data sheet, including appropriate hygienic practices, protective measures during repair and maintenance of contaminated equipment and procedures for clean-up of spills and leaks;

(ix) Any generally applicable control measures which are known to the chemical manufacturer, importer, or employer preparing the material safety data sheet, such as appropriate engineering control, work practices, or personal protective equipment;

(x) Emergency and first-aid procedures;

(xi) The date of preparation of the material safety data sheet or the last change to it; and,

(xii) The name, address and telephone number of the chemical manufacturer, importer or other responsible party preparing or distributing the material safety data sheet, who can provide additional information on the hazardous chemical and appropriate emergency procedures, if necessary.

(3) If no relevant information is found for any given category on the material safety data sheet, the chemical manufacturer, importer or employer preparing the material safety data sheet shall mark it to indicate that no applicable information was found.

(4) Where complex mixtures have similar hazards and contents (i.e., the chemical ingredients are essentially the same, but the specific composition varies from mixture to mixture), the chemical manufacturer, importer or employer may prepare a material safety data sheet to apply to all of these similar mixtures.

(5) The chemical manufacturer, importer or employer preparing the material safety data sheet shall ensure that the information recorded accurately reflects the scientific evidence used in making the hazard determination. If the chemical manufacturer, importer or employer preparing the material safety data sheet becomes newly aware of any significant information regarding the hazards of a chemical, or ways to protect against the hazards, this new information shall be added to the material safety data sheet within three months. If the chemical is not currently being produced or imported the chemical manufacturer or importer shall add the information to the material safety data sheet before the chemical is introduced into the workplace again.

(6) (i) The chemical manufacturer or importer shall ensure that distributors and employers are provided an appropriate material safety data sheet with their initial shipment, and with the first shipment after a material safety sheet is updated;

(ii) The chemical manufacturer or importer shall either provide material safety data sheets with the shipped containers or send them to the distributor or employer prior to or at the time of the shipment;

(iii) If the material safety data sheet is not provided with a shipment that has been labeled as a hazardous chemical, the distributor or employer shall obtain one from the chemical manufacturer or importer as soon as possible; and,

(iv) The chemical manufacturer or importer shall also provide distributors or employers with a material safety data sheet upon request.

(7) (i) Distributors shall ensure that material safety data sheets, and updated information, are provided to other distributors and employers with

their initial shipment and with the first shipment after a material safety data sheet is updated;

(ii) The distributor shall either provide material safety sheets with the shipped containers, or send them to the other distributor or employer prior to or at the time of the shipment;

(iii) Retail distributors selling hazardous chemicals to employers having a commercial account shall provide a material safety data sheet to such employers upon request, and shall post a sign or otherwise inform them that a material safety data sheet is available;

(iv) Wholesale distributors selling hazardous chemicals to employers over-the-counter my also provide material safety data sheets upon the request of the employer at the time of the over-the-counter purchase, and shall post a sign or otherwise inform such employers that a material safety data sheet is available;

(v) If an employer without a commercial account purchases a hazardous chemical from a retail distributor not required to have material safety data sheets on file (i.e., the retail distributor does not have commercial accounts and does not use the materials), the retail distributor shall provide the employer, upon request, with the name, address, and telephone number of the chemical manufacturer, importer or distributors from which a material safety data sheet can be obtained;

(vi) Wholesale distributors shall also provide material safety data sheets to employers or other distributors upon request; and,

(vii) Chemical manufacturers, importers or distributors need not provide material safety data sheets to retail distributors that have informed them that the retail distributor does not sell the product to commercial accounts or open the sealed container to use it in their own workplaces.

(8) The employer shall maintain in the workplace copies of the required material safety data sheets for each hazardous chemical, and shall ensure that they are readily accessible during each work shift to employees when they are in their work area(s). (Electronic access, microfiche, and other alternatives to maintaining paper copies of the material safety data sheets are permitted as long as no barriers to immediate employee access in each workplace are created by such options.)

(9) Where employees must travel between workplaces during a workshift, i.e., their work is carried out at more than one geographic location, the material safety data sheets may be kept at the primary workplace facility. In this situation, the employer shall ensure that employees can immediately obtain the required information in an emergency.

(10) Material safety data sheets may be kept in any form, including operating procedures, and may be designed to cover groups of hazardous chemicals in a work area where it may be more appropriate to address the hazards of a process rather than individual hazardous chemicals. However, the employer shall ensure that in all cases the required information is pro-

vided for each hazardous chemical, and is readily accessible during each work shift to employees when they are in their work area(s).

(11) Material safety data sheets shall also be made available, upon request, to designated representatives and to the Assistant Secretary, in accordance with the requirements of 29 CFR 1910.20(e). The Director shall also be given access to material safety data sheets in the same manner.

(h) "Employee Information and Training"

(1) Employers shall provide employees with effective information and training on hazardous chemicals in their work area at the time of their initial assignment, and whenever a new physical or health hazard the employees have not previously been trained about is introduced into their work area. Information and training may be designed to cover categories of hazards (e.g., flammability, carcinogenicity) or specific chemicals. Chemical-specific information must always be available through labels and material safety data sheets.

(2) "Information." Employees shall be informed of:
  (i) The requirements of this section;
  (ii) Any operations in their work area where hazardous chemicals are present; and,
  (iii) The location and availability of the written hazard communication program, including the required list(s) of hazardous chemicals, and material safety data sheets required by this section.

(3) "Training." Employee training shall include at least:
  (i) Methods and observations that may be used to detect the presence or release of a hazardous chemical in the work area (such as monitoring conducted by the employer, continuous monitoring devices, visual appearance or odor of hazardous chemicals when being released, etc.);
  (ii) The physical and health hazards of the chemicals in the work area;
  (iii) The measures employees can take to protect themselves from these hazards, including specific procedures the employer has implemented to protect employees from exposure to hazardous chemicals, such as appropriate work practices, emergency procedures, and personal protective equipment to be used; and,
  (iv) The details of the hazard communication program developed by the employer, including an explanation of the labeling system and the material safety data sheets, and how employees can obtain and use the appropriate hazard information.

(i) "Trade Secrets"

(1) The chemical manufacturer, importer or employer may withhold the specific identity, including the chemical name and other specific identification of a hazardous chemical, from the material safety data sheets, provided that:

(i) The claim that the information withheld is a trade secret can be supported;
(ii) Information contained in the material safety data sheets concerning the properties and effects of the hazardous chemical is disclosed;
(iii) The material safety data sheets indicate that the specific chemical identity is being withheld as a trade secret; and
(iv) The specific chemical identity is made available to health professionals, employees, and designated representatives in accordance with the applicable provisions of this paragraph.

(2) Where a treating physician or nurse determines that a medical emergency exists and the specific chemical identity of a hazardous chemical is necessary for emergency or first-aid treatment, the chemical manufacturer, importer, or employer shall immediately disclose the specific chemical identity of a trade secret chemical to that treating physical or nurse, regardless of the existence of a written statement of need or a confidentiality Agreement. The chemical manufacturer, importer, or employer may require a written statement of need and confidentiality agreement, in accordance with the provisions of paragraphs (i)(3) and (4) of this section, as soon as circumstances permit.

(3) In nonemergency situations, a chemical manufacturer, importer, or employer shall, upon request, disclose a specific chemical identity, otherwise permitted to be withheld under paragraph (i)(1) of this section, to a health professional (i.e., physician, industrial hygienist, toxicologist, epidemiologist, or occupational health nurse) providing medical or other occupational health services to exposed employee(s), and to employees or designated representatives, if:
  (i) The request is in writing;
  (ii) The request describes with reasonable detail one or more of the following occupational health needs for the information:
    (A) To access the hazards of the chemicals to which employees will be exposed;
    (B) To conduct or assess sampling of the workplace atmosphere to determine employee exposure levels;
    (C) To conduct preassignment or periodic medical surveillance of exposed employees;
    (D) To provide medical treatment to exposed employees;
    (E) To select or assess appropriate personal protective equipment for exposed employees;
    (F) To design or access engineering controls or other protective measures for exposed employees; and
    (G) To conduct studies to determine the health effects of exposure.
  (iii) The request explains in detail why the disclosure of the specific chemical identity is essential and that, in lieu thereof, the disclosure of the following information to the health professional, employee, or designated representative, would not satisfy the purposes described in paragraph (i) (3) (ii) of this section:

- (A) The properties and effects of the chemical;
- (B) Measures for controlling worker's exposure to the chemical;
- (C) Methods of monitoring and analyzing worker exposure to the chemical; and,
- (D) Methods of diagnosing and treating harmful exposures to the chemical;
  - (iv) The request includes a description of the procedures to be used to maintain the confidentiality of the disclosed information; and,
  - (v) The health professional, and the employer or contractor of the services of the health professional (i.e., downstream employer, labor organization, or individual employee), employee, or designated representative, agree in a written confidentiality agreement that the health professional, employee, or designated representative will not use the trade secret information for any purpose other than the health need(s) asserted and agree not to release the information under any circumstances other than to OSHA, as provided in paragraph (i) (6) of this section, except as authorized by the terms of the agreement or by the chemical manufacturer, importer, or employer.
- (4) The confidentiality agreement authorized by paragraph (i) (3) (iv) of this section:
  - (i) May restrict the use of the information to the health purposes indicated in the written statement of need;
  - (ii) May provide for appropriate legal remedies in the event of a breach of the agreement, including stipulation of a reasonable pre-estimate of likely damages; and,
  - (iii) May not include requirements for the posting of a penalty bond.
- (5) Nothing in this standard is meant to preclude the parties from pursuing noncontractual remedies to the extent permitted by law.
- (6) If the health professional, employee, or designated representative receiving the trade secret information decides that there is a need to disclose it to OSHA, the chemical manufacturer, importer or employer who provided the information shall be informed by the health professional, employee, or designated representative prior to, or at the same time as, such disclosure.
- (7) If the chemical manufacturer, importer or employer denies a written request for disclosure of a specific chemical identity, the denial must:
  - (i) Be provided to the health professional, employee, or designated representative, within thirty days of the request;
  - (ii) Be in writing;
  - (iii) Include evidence to support the claim that the specific chemical identity is a trade secret;
  - (iv) State the specific reasons why the request is being denied; and,
  - (v) Explain in detail how alternative information may satisfy the specific medical or occupational health need without revealing the specific chemical identity.
- (8) The health professional, employee, or designated representative whose

request for information is denied under paragraph (i) (3) of this section may refer the request and the written denial of the request to OSHA for consideration.

(9) When a health professional, employee, or designated representative refers the denial to OSHA under paragraph (i) (8) of this section, OSHA shall consider the evidence to determine if:

    (i) The chemical manufacturer, importer or employer has supported the claim that the specific chemical identity is a trade secret;

    (ii) The health professional, employee, or designated representative has supported the claim that there is a medical or occupational health need for the information; and

    (iii) The health professional, employee or designated representative has demonstrated adequate means to protect the confidentiality.

(10) (i) If OSHA determines that the specific chemical identity requested under paragraph (i) (3) of this section is not a "bona fide" trade secret, the employee, or designated representative, has a legitimate medical or occupational health need for the information, has executed a written confidentiality agreement, and has shown adequate means to protect the confidentiality of the information, the chemical manufacturer, importer or employer will be subject to citation by OSHA.

    (ii) If a chemical manufacturer, importer or employer demonstrates to OSHA that the execution of a confidentiality agreement would not provide sufficient protection against the potential harm from the unauthorized disclosure of a trade secret specific chemical identity, the Assistant Secretary may issue such orders or impose such additional information as may be appropriate to assure that the occupational health services are provided without an undue risk of harm to the chemical manufacturer, importer or employer.

(11) If a citation for a failure to release specific identity information is contested by the chemical manufacturer, importer or employer, the matter will be adjudicated before the Occupational Safety and Health Review Commission in accordance with the Act's enforcement scheme and the applicable Commission rules of procedure. In accordance with the Commission rules, when a chemical manufacturer, importer or employer continues to withhold the information during the contest, the Administrative Law Judge may review the citation and supporting documentation "in camera" or issue appropriate orders to protect the confidentiality of such matters.

(12) Notwithstanding the existence of a trade secret claim, a chemical manufacturer, importer, or employer shall, upon request, disclose to the Assistant Secretary any information which this section requires the chemical manufacturer, importer, or employer to make available. Where there is a trade secret claim, such claim shall be made no later than at the time the information is provided to the Assistant Secretary so that

suitable determinations of trade secret status can be made and the necessary protections can be implemented.

(13) Nothing in this paragraph shall be construed as requiring the disclosure under any circumstances of process or percentage of mixture information which is a trade secret.

(j) "Effective Dates"

Chemical manufacturers, importers, distributors, and employers shall be in compliance with all provisions of this section by March 11, 1994.

Note: The effective date of the clarification that the exemption of wood and wood products from the Hazard Communication Standard in paragraph (b) (6) (iv) only applies to wood and wood products including lumber which will not be processed, where the manufacturer or importer can establish that the only hazard they pose to employees is the potential for flammability or combustibility, and that the exemption does not apply to wood or wood products which have been treated with a hazardous chemical covered by this standard, and wood which may be subsequently sawed or cut generating dust has been stayed from March 11, 1994 to August 11, 1994.

Appendix D

Action Items Control Log

The action items control log (AICL) is the single source document used to manage the constant flow of requirements for actions to resolve issues necessary to complete the project. This log is computer generated and maintained by the project engineer. Copies are distributed on a regular basis for use by all project participants. The log is reviewed at all project progress meetings and other meetings as required.

The form enclosed with this appendix is an example of an AICL. The procedure for use of the log follows, including a description of each column in the log and the procedure for the log's use.

Procedure

Number. Sequential number used to keep a tally of the total number of action items. This number is one greater than the previous action item number. It is assigned by the person who maintains the log.

Item type. Types of correspondence should be identified by an alpha abbreviation. Types and identifiers are as follows:

RE	Resident engineer's letter
PE	Project engineer's letter
LTR	Letter
RFI	Request for information
RD	Resident's directive
EWO	Extra work order
CO	Change order
SCH	Schedule
S-DWG	Shop drawing
R-DWG	Revised drawing
XMTL	Transmittal

PL	Punchlist
MTG	Meeting
HAZ	Items related to the discovery of hazardous materials

Date. Date the sender placed on the correspondence.

From. Identity of the sender. The name of the company or firm is sufficient.

Date received. Date corresponding to the "received date" stamped on the incoming correspondence.

Action required. A brief but complete description of the action required by the addressee. Since one item of correspondence may contain more than one action item, list each action item as a subitem of the main item. For example, if item number 6 on the log contains three separate action items, list the items as 6a, 6b, and 6c. Then, complete columns 7, 8, and 9 for each subitem.

Action party. Identify not only the agency, but also the individual responsible for closing out the action item. The responsible party is also responsible for ensuring that completion or extension dates are properly recorded in the project action items control log.

Date due. Date the identified action is due to be completed. It is the responsibility of the action party to ensure that the due date is met.

Date closed. The actual date on which the action item is completed.

Closed with. Record the letter number, transmittal number, or other description of document used to close the action.

The project engineer maintains the project action items control log for the project.

The project action items control log is reviewed during each periodic project meeting. The project engineer distributes copies of the log to all responsible parties noted on the log at each meeting.

NO.	ITEM TYPE	DATE	FROM	DATE RECEIVED	ACTION REQUIRED	ACTION PARTY	DATE DUE	DATE CLOSED	CLOSED WITH

ACTION ITEMS CONTROL LOG

Appendix E

Sample Written Hazard Communication Program

Note: This model program is provided by OSHA only as a guideline to assist in complying with 29 CFR 1910.1200. It is not intended to supersede the requirements of 29 CFR 1910.1200. Employers should review the Hazard Communication Standard for particular requirements which are applicable to the specific situation.

Sample Written Hazard Communication Program

1. Company Policy

   To insure that information about the dangers of all hazardous chemicals used by __(name of company)__ are known by all affected employees, the following hazardous information program has been established:

   All work units of the company will participate in the hazard communication program. This written program will be available in the __(location)__ for review by any interested employee.

2. Container Labeling

   The __(person/position)__ will verify that all containers received for use will be clearly labeled as to the contents, note the appropriate hazard warning and list the name and address of the manufacturer.

   The __(person/position)__ in each section will ensure that all secondary containers are labeled with either an extra copy of the original manufacturer's label or with labels that have the identity and the appropriate hazard warning. For help with labeling, see __(person/position)__.

   On the following individual stationary process containers, we are using

_____ rather than a label to convey the following required information:

_____

We are using an in-house labeling system which relies on:

(Please provide a description of any in-house system which uses numbers or graphics)

_____
_____
_____

The __(person/position)__ will review the company labeling procedures every __(time period)__ and update as required.

3. Material Safety Data Sheets (MSDSs)

The __(person/position)__ is responsible for establishing and monitoring the company MSDS program. He/she will make sure procedures are developed to obtain the necessary MSDSs and will review incoming MSDSs for new or significant health and safety information. He/she will see that any new information is passed on to affected employees. The following procedure will be followed when an MSDS is not received at the time of initial shipment:

_____
_____

Copies of MSDSs for all hazardous chemicals in use will be kept in __(location)__.

MSDSs will be readily available to all employees during each work shift. If an MSDS is not available, immediately contact __(person/position)__. To ensure MSDSs are readily available in each work area, the following format will be used:

_____

Note: If an alternative to material safety data sheets is used, provide a description for the format. When revised MSDSs are received, the following procedures will be followed to replace old MSDSs: _____

_____

4. Employee Training and Information

The __(person/position)__ is responsible for the company employee training program. He/she will ensure that all program elements specified below are carried out.

Prior to starting work, each new employee of __(name of company)__ will attend a health and safety orientation that includes the following information and training:

- An overview of the requirements contained in the Hazard Communication Standard.
- Hazardous chemicals present at his/her workplaces.
- Physical and health risks of the hazardous chemicals.
- The symptoms of overexposure.
- How to determine the presence or release of hazardous chemicals in his/her work area.
- How to reduce or prevent exposure to hazardous chemicals through use of control procedures, work practices, and personal protective equipment.
- Steps the company has taken to reduce or prevent exposure to hazardous chemicals.
- Procedures to follow if employees are overexposed to hazardous chemicals.
- Location of the MSDS file and written hazard communication program.

Prior to introducing a new chemical hazard into any section of this company, each employee in that section will be given information and training as outlined above for the new chemical hazard. The training format will be as follows: __(audio-visuals, classroom instructions, etc.)__ .

5. Hazardous Nonroutine Tasks

    Periodically, employees are required to perform hazardous nonroutine tasks. Some examples of nonroutine tasks are: confined space entry, tank cleaning, and painting reactor vessels. Prior to starting work on such projects, each affected employee will be given a list of hazardous chemicals he or she may encounter during such activity. This information will include specific chemical hazards, protective and safety measures the employees can use, and steps the company is using to reduce the hazards, including ventilation, respirators, presence of another employee and emergency procedures.

    Examples of nonroutine tasks performed by employees of this company:

Tasks	Hazardous Chemical

6. Informing Other Employers

    It is the responsibility of __(person/position)__ to provide other employers with information about hazardous chemicals their employees may be exposed to on a job site and suggested precautions for the employees. It is the responsibility of __(person/position)__ to obtain information about hazardous chemicals used by other employers to which employees of this company may be exposed.

    (The following option is recommended by OSHA.)

Other employers will be contracted before work is started to gather and distribute information concerning any chemical hazard that may be present in the workplace.

7. List of Hazardous Chemicals

The following is a list of all known hazardous chemicals used by our employees. Further information on each chemical may be obtained by reviewing the MSDSs located at __(location)__ .

MSDS Identity

(Include the chemical list developed during the inventory. Arrange this list so that you are able to cross-reference it with your MSDS file and the labels on your containers.)

Chemicals not already on the list are added to the list (together with dates the chemicals were introduced), within 30 days of introduction into the workplace. To ensure that the chemical is added in a timely manner, we have introduced the following procedures: _____

_____

Our chemical information list was compiled and is maintained by: __(name and telephone number of responsible person)__

8. Chemicals in Unlabeled Pipes

Work activities are sometimes performed by employees in areas where chemicals are transferred through unlabeled pipes.

Prior to starting work in these areas, the employee shall contact __(person/position)__ for information regarding:

- The chemical in the pipes.
- Potential hazards.
- Safety precautions which should be taken.

9. A copy of this program will be made available, upon request, to employees and their representatives.

10. The following methods will be used to inform other employers who have employees who may be exposed to hazardous chemicals used by employees of this company:

   a) Material safety data sheets will be provided to other employers in the following manner:_____

   b) Appropriate precautionary methods will be related to other employers to safeguard their employees.

   c) Other employers will be informed of the labeling system in use.

Notes for Chemical Manufacturers, Importers, and Distributors

1. Hazard Determination. Chemical manufacturers and importers are to detail the methods they will use to conduct a hazard determination for chemicals produced or imported in their work places. The procedures should identify the system in place to conduct hazard determinations. The system should identify the person or department responsible for conducting the determination and the research strategy involved. Chemical manufacturers which rely on information from upstream chemical manufacturers should state this in their written program.

2. Transmittal of Material Safety Data Sheets. Chemical manufacturers, importers, and distributors should develop a system to ensure that material safety data sheets are transmitted to customers. The system should identify the person of the department responsible for ensuring the transmittal of material safety data sheets and should include a method to ensure that transmittal is accomplished as required by 29 CFR 1910.1200.

3. Labels. Chemical manufacturers, importers, and distributors should have a system for ensuring appropriate labeling of hazardous chemicals.

4. Updating Labels/MSDSs. A system should be detailed assigning responsibility and periodic review of scientific information required to update material safety data sheets and labels as required by 29 CFR 1910.1200.

Hazard Communication Checklist

_____ 1. Has a list been prepared of all hazardous chemicals in the workplace?

_____ 2. Is the company prepared to update the hazardous chemical list?

_____ 3. Has the company obtained or developed a material safety sheet for each hazardous chemical we use?

_____ 4. Has a system been developed to ensure that all incoming hazardous chemicals are checked for proper labels and data sheets?

_____ 5. Are procedures in place to ensure labeling or warning signs for containers that hold hazardous chemicals?

_____ 6. Are employees aware of the specific information and training requirements of the Hazard Communication Standard?

_____ 7. Are employees familiar with different types of chemicals and hazards associated with them?

_____ 8. Have employees been informed of the hazards associated with performing nonroutine tasks?

_____ 9. Do employees understand how to detect the presence of release of hazardous chemicals in the workplace?

_____10. Are employees trained about proper work practices and personal protective equipment in relation to the hazardous chemicals in their work area?

_____11. Does the training program provide information on appropriate first aid, emergency procedures, and the likely symptoms of over exposures?

____12. Does the training program include an explanation of labels and warnings that are used in each work area?

____13. Does the training describe where to obtain data sheets and how employees may use them?

____14. Is a system in place to ensure that new employees are trained before beginning work?

____15. Is a system in place to identify new hazardous chemicals before they are introduced into a work area?

____16. Is a system in place to inform employees of a new hazard associated with a chemical?

Appendix F

Excerpt from *Modification Impact Evaluation Guide**

---

*Department of the Army, Office of the Chief of Engineers, EP 415-1-3, July 2, 1979, pp. 4-6 to 4-17.

EP 415-1-3
2 July 79

(1) <u>Disruption</u>. The contractor's progress schedule represents the planned sequence of activities leading to final completion of the project. Workers who know what they are doing, what they will be doing next, and how their activities relate to the successful completion of the project develop a "job rhythm." Labor productivity is at its optimum when there is good job rhythm. When job rhythm is interrupted (i.e., when a contract modification necessitates a revision of the progress schedule), it affects workers on both the directly changed and/or unchanged work and may result in a loss of productivity.

(a) Disruption occurs when workers are prematurely moved from one assigned task to another. Regardless of the competency of the workers involved, some loss in productivity is inevitable during a period of orientation to a new assignment. This loss is repeated if workers are later returned to their original job assignment. Learning curves which graph the relationship between production rate and repeated performance of the same task have been developed for various industrial tasks. The basic principle of all learning curve studies is that efficiency increases as an individual or team repeats an operation over and over; assembly lines are excellent demonstrations of this principle. However, although construction work involves the repetition of similar or related tasks, these tasks are seldom identical. Skilled construction workers are trained to perform a wide variety of tasks related to their specific trade. Therefore, in construction it is more appropriate to consider the time required to become oriented to the task rather than acquiring the skill necessary to perform it. One of the attributes of the construction worker is the ability to perform the duties of his trade in a variety of environments. How long it will take the worker to adjust to a new task and environment depends on how closely related the task is to his experience or how typical it is to work usually performed by his craft. Figure 4-1a assumes that the worker will always be assigned to perform work within the scope of his trade, and that the average worker will require a maximum of one shift (8 hours) to reach full productivity. Full productivity (100 on the Theoretical Productivity Scale) represents optimum productivity for a given project. Figure 4-1b is a tabulation of productivity losses derived from figure 4-1a.

(b) The time required for a worker (or crew) to reach full productivity in a new assignment is not constant. It will vary with skill, experience, and the difference between the old and new task. In using the chart or its tabulation, the estimator must decide what point on the Theoretical Productivity Scale represents a composite of these factors. For example, an ironworker is moved from placing reinforcing bars to the structural steel erection crew. The ironworker is qualified by past training to work on structural steel, but the vast majority of his

EP 415-1-3
2 July 79

experience has been with rebars, and the two tasks are significantly different. In view of this, a starting point of "0" is appropriate. The estimator can determine from the chart that a "0" starting point indicates the ironworker will need 8 hours to reach full productivity, with a resulting productivity loss of 4 hours. The Government's liability is then 4 hours times the hourly rate times markups. As a second example, assume the same ironworker is moved from placing reinforcing bars for Building A to placing reinforcing bars for Building B. The buildings are similar but not identical. A starting point of "90" is appropriate. The duration of only 0.8 hours is required to reach full productivity, and the productivity loss is 0.4 hours. The Government's liability would then be 0.4 hours times the hourly rate times markups.

(c) The contractor normally absorbs many orientation/learning cycles as his labor forces are moved from task to task in the performance of the work. Only those additional manpower moves, caused solely by a contract modification, represent labor disruption costs for which the contractor is entitled extra payment.

(2) <u>Crowding</u>. If a contractor's progress schedule is altered so that more activities must be accomplished concurrently, impact costs caused by crowding can result. Crowding occurs when more workers are placed in a given area than can function effectively. Crowding causes lowered productivity; it can be considered a form of acceleration because it requires the contractor either to accomplish a fixed amount of work within a shorter time frame, or to accomplish more work within a fixed time frame. Granting additional time for completion of the project can eliminate crowding. When the final completion date cannot be slipped, increased stacking of activities must be analyzed and quantified.

(a) Activity stacking does not necessarily result in crowding -- when concurrent activities are performed in areas where working room is sufficient, crowding is not a factor. But, if the modification forces the contractor to schedule more activities concurrently in a limited working space, crowding does result. Both increased activity stacking and limited (congested) working space must be present for crowding to become an item of impact cost.

(b) Crowding can be quantified by using techniques similar to those used for acceleration. Figure 4-2 illustrates the curve developed to represent increases in labor costs from crowding. Before applying this curve, the estimator must determine whether crowding will occur and to what degree. For example, the assumption that the contractor's scheduling of the activities in question is the most efficient

EP 415-1-3
2 July 79

sequencing of the work must be verified. Perhaps more workers can work effectively in the applicable work space than the contractor has scheduled; if they cannot, perhaps the crowding is not severe enough to justify using the full percentage of loss indicated by the graph. (The graph should be interpreted as representing the upper limit of productivity loss.) In this case, the estimator's judgment of the specific circumstances may indicate that some lower increase factor is appropriate.

(c) For example, assume that the estimator decides that severe crowding will occur in the following situation: The contractor's schedule indicates three activities concurrently in progress in a limited area of the project. Each of these activities employs five workers, placing a total of 15 workers in the area. One of these activities has a duration of 10 days; the other two have 20-day durations. The modification has required that a fourth activity be scheduled concurrently in the same limited area. This additional activity requires three workers; it has a normal duration of 5 days. There are now 18 workers in an area which can only efficiently accommodate 15. The percent of crowding is 3/15 or 20 percent. On the graph (figure 4-2), 20 percent crowding intersects the curve opposite 8 percent loss of efficiency. To find the duration of crowding, the estimator multiplies the normal duration of the added activity by 100 percent plus the percent loss of efficiency. For this example, 5 days times 1.08 equals 5.4 days. Therefore, because of the inefficiency introduced by crowding, the added activity will require 5.4 days to complete. Likewise, on the three affected activities, the first 5 days of normal activity will now require 5.4 days. All four activities will experience loss of productivity resulting from an inefficiency factor equivalent to 0.4 of a single day's labor cost. This is calculated as follows:

Average hourly rate x hours worked per day x number of workers x 0.4 = $ loss

or

$12.00 x 8 x 18 x 0.4 = $691 plus normal labor markups.

3/18 x $691 = direct crowding cost; and should be included in the Direct Cost section of the modification estimate

15/18 x $691 = crowding on unchanged activities, and should be placed in the Impact on Unchanged Work section of the modification estimate.

EP 415-1-3
2 Jul 79

(3) <u>Acceleration</u>. Acceleration occurs when a modification requires the contractor to accomplish a greater amount of work during the same time period even though he may be entitled to an extension of time to accomplish the changed work. This is sometimes referred to as "buying back time." Acceleration should be distinguished from expediting. Expediting occurs whenever the modification would require the contractor to complete the work before the original completion date included in the contract. Per DAR 18-111, expediting is not permissible in the absence of approval by the Assistant Secretary of Defense (Manpower, Reserve Affairs and Logistics). Acceleration may be accomplished in any of the following ways:

(a) <u>Increasing the size of crews</u>. The optimum crew size (for any construction operation) is the minimum number of workers required to perform the task within the allocated time frame. Optimum crew size for a project or activity represents a balance between an acceptable rate of progress and the maximum return from the labor dollars invested. Increasing crew size above optimum can usually produce a higher rate of progress, but at a higher unit cost. As more workers are added to the optimum crew, each new worker will increase crew productivity less than the previously added worker. Carried to the extreme, adding more workers will contribute nothing to overall crew productivity. Figures 4-3a through 4-3d indicate the effect of crew overloading.

(b) <u>Increasing shift length and/or days worked per week</u>. The standard work week is 8 hours per day, 5 days per week (Monday through Friday). Working more hours per day or more days per week introduces premium pay rates and efficiency losses. Workers tend to pace themselves for longer shifts and more days per week. An individual or a crew working 10 hours a day, 5 days a week, will not produce 25 percent more than they would working 8 hours a day, 5 days a week. Longer shifts will produce some gain in production, but it will be at a higher unit cost than normal hour work. When modifications make it necessary for the contractor to resort to overtime work, some of the labor costs produce no return because of inefficiency. Costs incurred due to loss of efficiency created by overtime work are an impact element because the increase in overtime results from the introduction of the modification. Contractors occasionally find that to attract sufficient manpower and skilled craftsmen to the job, it is necessary to offer overtime work as an incentive. When this is done, the cost must be borne by the contractor; however, if overtime is necessary to accomplish modification work, the Government must recognize its liability for introducing efficiency losses. Figure 4-4 is the result of study which attempted to graphically demonstrate efficiency losses over a 4-week period for several combinations of work schedules. These data are included merely

EP 415-1-3
2 Jul 79

as information on trends rather than firm rules which might apply to any project. Although figure 4-4 data do not extend beyond the fourth week, it is assumed that the curves would flatten to a constant efficiency level as each work schedule is continued for longer periods of time.

(c) <u>Multiple shifts</u>. The inefficiencies in labor productivity caused by overtime work can be avoided by working two or three 8-hour shifts per day. However, additional shifts introduce other costs. These costs would include additional administrative personnel, supervision, quality control, lighting, etc. Modifications that cause the contractor to implement shift work should price the impact cost as appropriate for the activity being accelerated. Environmental conditions such as lighting and cold weather may also influence labor efficiency.

(4) <u>Morale</u>. The responsibility for motivating the work force and providing a psychological environment conducive to optimum productivity rests with the contractor. Morale does exert an influence on productivity, but so many factors interact on morale that their individual effects defy quantification. A project's contract modifications, particularly a large number, have an adverse effect on the morale of the workers. The degree to which this may affect productivity, and consequently the cost of performing the work, would normally be very minor when compared to the other causes of productivity loss. A contractor would probably find that it would cost more to maintain the records necessary to document productivity losses from lowered morale than justified by the amount he might recover. Modification estimates do not consider morale as a factor because whether morale becomes a factor is determined by how effective the contractor is in his labor relations responsibilities.

4-5. <u>Quantification</u>. The following example demonstrates how to use figures 4-3a through 4-3d to quantify the impact costs of crew overloading. Assume that the contractor has planned a construction operation with a duration of 15 working days and an optimum crew size of 10. The modification now requires that the contractor accomplish this operation in 10 working days. The rate of production is the unit of work per amount of effort in mandays. The percent increase is new rate minus original rate divided by original rate times 100. Thus,

$$\frac{(1 \text{ job} \div 100 \text{ MD}) - (1 \text{ job} \div 150 \text{ MD})}{1 \text{ job} \div 150 \text{ MD}} \times 100 =$$

$$\frac{.01 - .0067}{.0067} \times 100 = 50 \text{ percent}$$

4-10

EP 415-1-3
2 Jul 79

This represents a 50 percent increase in the crew's rate of production. From figure 4-3a or 4-3d, it appears likely that 50 percent production gain can be achieved by increasing the size 80 percent. Other options could be implemented to speed up production: the optimum crew could work longer shifts, more days per week; a second crew could be placed in operation (if allowed by the nature of the work). However, for this example only increasing crew size is considered. The way to quantify the impact cost before the fact is:

	Original Plan	Accelerated Plan
Manpower	10	18
Hourly Rate	$12	$12
Crew Cost/Day (8 hours)	$960	$1,728
Duration (Working Days)	15	10
Crew Cost (Cost/Day x Duration)	$14,400	$17,280
Taxes, Insurance, Fringes (18 percent)	$2,595	$3,110
Total Crew Cost	$16,992	$20,390

Impact Cost (Accelerated-Original) = $3,398 ($3,400)

- or -

Impact Cost (Accelerated Plan x Efficiency Loss) =
    $20,390 x 16.7 percent (from fig. 4-3b), = $3,405 ($3,400)

The amount of $3,400 would be placed in the modification estimate, under "Impact on Unchanged Work" and identified by the activity involved. Increased cost of supervision, if necessary, is not included in this crew overloading analysis. Supervision must be costed separately, either as a separate item or as an element of Job Site Overhead, as appropriate.

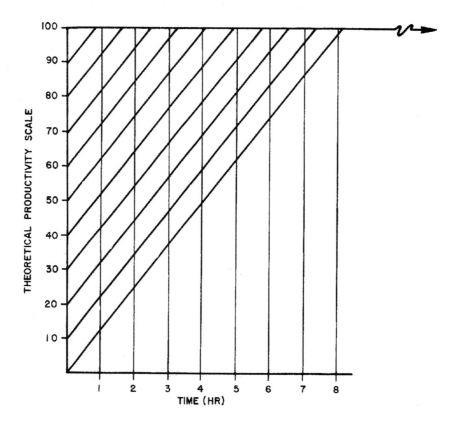

*100 REPRESENTS THE PRODUCTIVITY RATE REQUIRED TO MAINTAIN SCHEDULED PROGRESS

Figure 4-1a. Construction operations orientation/learning chart.

Excerpt from *Modification Impact Evaluation Guide*

EP 415-1-3
2 July 79

(BASED ON CONSTRUCTION OPERATIONS
ORIENTATION/LEARNING CHART)

PRODUCTIVITY STARTING POINT	DURATION (HR)	AVERAGE LOSS (HR)
100	0	0
90	0.8	0.4
80	1.6	0.8
70	2.4	1.2
60	3.2	1.6
50	4.0	2.0
40	4.8	2.4
30	5.6	2.8
20	6.4	3.2
10	7.2	3.6
0	8.0	4.0

Figure 4-1b. Productivity losses derived from figure 4-1.1.

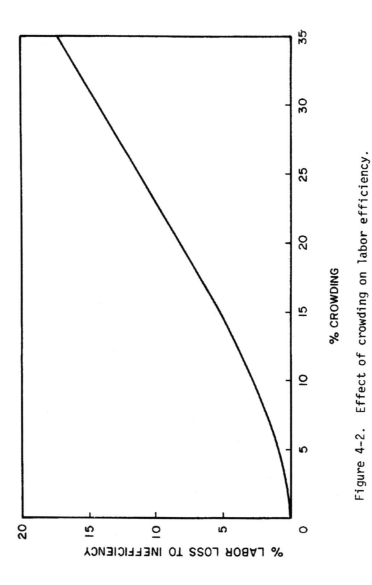

Figure 4-2. Effect of crowding on labor efficiency.

Excerpt from *Modification Impact Evaluation Guide* 247

EP 415-1-3
2 July 79

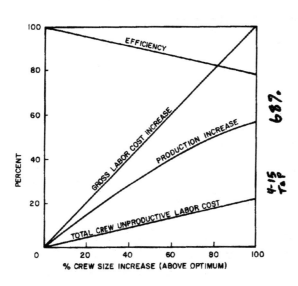

Figure 4-3 a. Composite effects of crew overloading

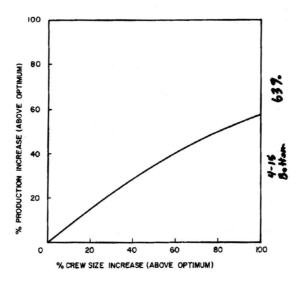

Figure 4-3 b. Unproductive labor at crew overloading.

4-15

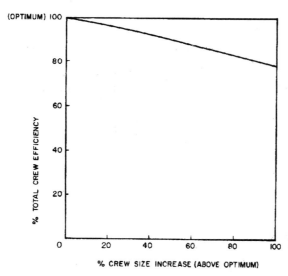

Figure 4-3c. Efficiency of crew overloading.

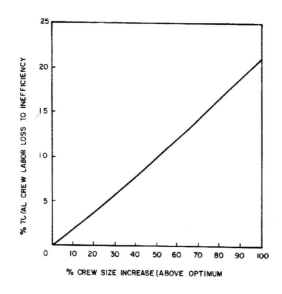

Figure 4-3d. Production gain of crew overloading.

Excerpt from *Modification Impact Evaluation Guide* 249

EP 415-1-3
2 July 79

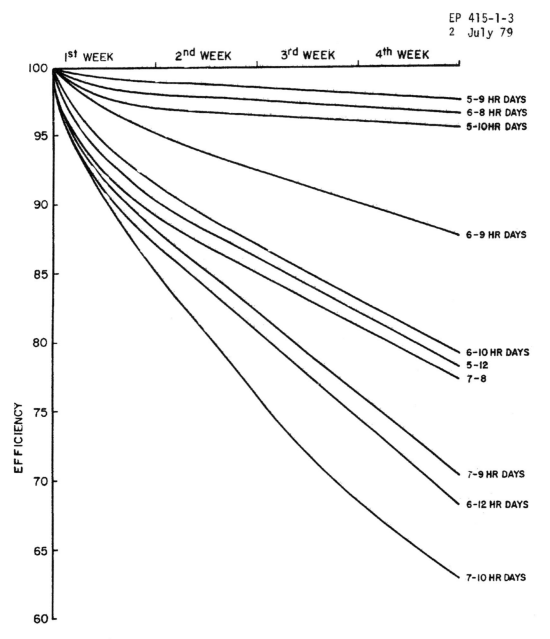

Figure 4-4. Effect of work schedule on efficiency.

4-17

Index

Acceleration, 137–139, 148–149
    cost increases attributable to, 149
Action items control log, 47, 227–229
Action items list, 116–117
Activity-on-arrow (AOA) format, 58
Activity-on-node (AON) format, 58
Actual costs, 151
Additional equipment costs, 159–161
    owned equipment rates, 160–161
    rented equipment rates, 159–160
Additional labor, costs of, 152–153
Additional material costs, 164
AED Green Book, 160–161
Alternative Dispute Resolution (ADR) (*see* Nonbinding dispute resolution)
American Arbitration Association (AAA), 181–182
American Society of Home Inspectors (ASHI), 13
As-built CPM schedule, 131–132
Asbestos, 1–3
    definition of, 2
    fibers, 1
    removal of, 2–3
Assigning risk, 28–33

Bar chart, 56–58
    for analysis of delays, 127–129
    example of, 57
Binding arbitration, 181–182
Bonds, 40–41, 172
    bid bond, 40
    payment bond, 40
    performance bond, 40
        example of, 42
Bonus/penalty provision, 86

Canadian method (for calculating a daily rate for home office overhead), 169
Cardinal change, 74
    definition of, 74
Certification, 171
    example of, 171
Change order, 73–74

Change Order Proposal (COP), 52
Change Order Request (COR), 52
Changes clause, 68–79
    changed conditions, 69
    concealed conditions, 69
    differing site conditions, 69–72
    elements of, 68
Collapsed as-built approach, 131
Communication within the project team, 64–66
Comprehensive Environmental Response, Compensation, and Liability Act (CERCLA), 1, 8
Constructability review, 40
Construction Equipment Ownership and Operating Expense Schedule, 161
Construction manager (CM), 25–27, 33
Constructive acceleration, 138–139
Containment, of contamination, 111–112
Contaminant(s):
    asbestos, 1
    cadmium, 1
    definition of, 1
    lead, 1
    polychlorinated biphenyls (PCBs), 1
Contamination threat assessment, 22
Contemporaneous analysis approach, 131–136
Continuation-of-work clause, 74
    example of, 74
Contract appeals boards, 181–182
Contract forms, 67–68
Contract notice provision, 64–66, 72–73
Contract relationship, 25–28
Contractor's cost record clause, 153
Controlling delay (*see* Critical delay)
Cost effect, 151
Cost estimate, 151
Cost of cleanup, liability, 8–9
Cost-plus contract, 31–33, 147
Cost-plus work, 75
Critical delay, 121–122
Critical path, 62–63
    definition of, 62
Critical path method (CPM) schedule, 58–64, 80–84, 121, 136

251

Critical path method (CPM) schedule (*Cont.*):
for analysis of delays, 122–127
example of, 59–62
Current owner, definition of, 10

Default termination, 92
clause, 92–95
examples of, 92–94
Delay, 121
Design/build:
project, 33
team, 27–28
Design review, 39–40
Dilution of supervision, 143
Direct cost, of labor, 152
Directed acceleration, 138
Discoverable document, 114
Disputes review board (DRB), 54, 96–99, 179–180
Document organization, 117–118
Documentation (of discovery of contamination), 113–119
daily reports, logs, and diaries, 114
letters and memoranda, 114–116
meeting minutes and other project documentation, 116–117
other, 118–119

Efficiency, 141–142, 145
(*See also* Productivity)
Eichleay formula, 168–169
Environmental audit, 9
Environmental Protection Agency (EPA), 2, 105, 110
Equipment cost effects, 157–163
additional equipment costs, 159–161
owned equipment rates, 160–161
rented equipment rates, 159–160
equipment cost escalation, 162–163
extended or idle equipment costs, 161–162
Escrow of bid documentation, 95–96
Estimated costs, 151
Evaluation (of contamination), 112–113
Exculpatory clause, 29–30
Execution, 39–66
Extended equipment:
costs associated with, 161–162
definition of, 161
Extended labor costs, 153–154
idle labor costs, 153–154
Facilitator, in partnering session, 44–45

Fast-tracking, fast-tracked projects, 34–35, 37
Field overhead costs, 165, 167
definition of, 165
field office overhead costs, 168
Fixed fee, 32
Fixed-price contract, 147
form (*see* Lump-sum contract, form)
Force account, 75
Fragnet, 124–127
Fuels, in storage tanks, 4

Gantt chart (*see* Bar chart)
General contractor, 27
Guaranteed maximum price (GMP), 25, 32–33

Hazard communication checklist, 235–236
Hazard Communication Program (HCP), 109–110
contents, 109–110
sample written, 231–236
Hazard Communications Standard (HCS), 105–106, 110
OSHA regulations, 205–225
definitions, 209–213
effective dates, 225
employee information and training, 221
hazard determination, 213–215
labels and other forms of warning, 216–217
material safety data sheets, 217–221
purpose, 205–206
trade secrets, 221–225
Hazardous material, definition of, 1
Hazardous waste, definition of, 1
Home office overhead, 165
costs, 168
definition of, 165
Housing and Community Development Act, 3

Idle labor costs, 153–154
Idle time:
costs associated with, 162
definition of, 161–162
"If and where directed" item contract clause, 31
example of, 32
Impacted as-planned analysis, 129–131
Incentive/disincentive provision, 86
example of, 86

Innocent purchaser, 8
Interest, 170

Labor cost effects:
  additional labor, 152–153
  definition of, 152
  extended and idle labor costs, 153–154
  labor escalation, 154–157
Lead, 1, 3–4
  hazards of, 3
Lead-based paints, 3–4, 38
  removal, 4
Liquidated damages, 85–86
  definition of, 85
Logic network (*see* Critical path method schedule)
Lump-sum contract, 33
  form, 31

Material cost effects, 164
  additional material costs, 164
  material cost escalation, 164
  material storage costs, 164
Material cost escalation, 164
Material safety data sheet (MSDS), 106–110
  example of, 107–108
Material storage costs, 164
Measured mile analysis, 145–147
Mediation, 177–179
Mission statement, in partnering charters, 47
Modification Impact Evaluation Guide, excerpt from, 237–249
Modification of the contract, 73–74
Monitoring responsibilities, 52–55

Narrative report, example of, 65
Negotiation, 174–177
  preparation, 176
  time limits, 175
No-damage-for-delay clause, 29–30, 170
Nonbinding arbitration, 181
Nonbinding dispute resolution, 173–181
  disputes review boards, 179–180
  mediation, 177–179
  minitrials, 180
  negotiation, 174–177
  nonbinding arbitration, 181
  use of settlement judge or court-appointed master, 179–180
Noncontrolling delay (*see* Noncritical delay)
Noncritical delay, 121–122

Not-to-exceed price, 32
  (*See also* Guaranteed maximum price)
Notification letter, 103–104
  example of, 104

Occupational Safety and Health Administration (OSHA), 2, 105–106, 110, 120
  general requirements regarding hazardous waste operations, summary of, 110–111
Overhead, 165–170

Partnering, 43–47
  charters, examples of, 45–46
  contract provisions for, 43–44
  definition of, 43
  sessions, 43–45
Payment, for change work, 74–79
Payroll burden, 152–153
Planning efforts, 21–39
  contractor, 38–39
  designer, 37–38
  owner, 22
Polychlorinated biphenyls (PCBs), 1, 4–5
  definition of, 4–5
Pre-bid meeting, 41–43
Prequalification (of contractor):
  criteria for, 196–198
  grounds for denial, 198–199
Prequalification Process Questionnaire, 199–204
Prequalification Questionnaire:
  instructions for completing, 199
  Special Contractor, 25, 195–204
Productivity, 141–143, 145, 147, 149
  (*See also* Efficiency)
Profit, 170
Program manager (PM), 36, 63
Project coordinator (PC) (*see* Construction manager)
Project management, 21
Project manager (PM) (*see* Construction manager)
Project manual, 47
  components of, 48
Project team, selection of, 23–25
Project timing, 34–37

Rental Rate Blue Book, 160–161
Request for Information (RFI), 48–50
  example of, 49
Request for Proposal (RFP), 15, 23

Requirements for Owners and Operators of Underground Storage Tanks, 183–194
Resolution, steps of, 53–55
Resource Conservation and Recovery Act (RCRA), 1, 105
Response time, to questions, submittals, and problems of contractor, 47–52
Risk-shifting clause, 18
Role in execution of project:
  contractor, 55–66
  owner, 39–55

Safety and health response (to contamination), 105–111
Scheduling, 56–64
Site inspection clause, 17
  definition of, 17
Standard of normal conditions, 71
Standard operating procedure (SOP), 105
Submittal form, example of, 51
Suspensions of work, 84–85
  contract clause, 84

Termination for cause (*see* Default termination)

Termination for convenience:
  contract clause, 21–22, 89–95
  example of, 89–92
Termination of a contract, 55
Testing firm, selection of, 14
Time-and-materials contract (*see* Cost-plus contract)
Time extensions clause, 86–88
  example of, 86–88
Total cost approach, 149
Total cost method, 147–148
Training crew to recognize contamination, 102–103

Underground storage tanks (USTs), 4
  requirements for owners and operators of, 183–194
Unit-price contract, 31
  (*See also* Fixed-price contract)

Variation in quantities clause, 80
  example of, 80

Wrongful termination, 94

ABOUT THE AUTHORS

J. Scott Lowe, P.E., is senior vice-president of Trauner Consulting Services, Inc. in Philadelphia. His expertise is in the areas of dispute resolution, construction management, CPM scheduling, contract administration, and cost analysis. He is a civil engineering graduate of Northwestern University.

Theodore J. Trauner, Jr., P.E., P.P., is principal and CEO of Trauner Consulting and the author of numerous construction books and professional articles. Mr. Trauner is a West Point graduate and holds a Masters of Engineering degree from the University of California at Berkeley.